7/5/18

GARDENING UNDER LIGHTS

GARDENING

TIMBER PRESS · PORTLAND, OREGON

UNDER LIGHTS

The Complete Guide for Indoor Growers

LESLIE F. HALLECK

CONTENTS

Published in 2018 by Timber Press, Inc.
The Haseltine Building
133 S.W. Second Avenue, Suite 450
Portland, Oregon 97204-3527
timberpress.com

Printed in China
Text and cover design by Adrianna Sutton

Library of Congress Cataloging-in-Publication Data

Names: Halleck, Leslie F., author.
Title: Gardening under lights: the complete guide for indoor growers /
    Leslie F. Halleck.
Description: Portland, Oregon: Timber Press, 2018. | Includes bibliographical
    references and index. | Identifiers: LCCN 2018001914 (print) | LCCN 2018004015
    (ebook) | ISBN 9781604698657 | ISBN 9781604697957 (hardcover)
Subjects: LCSH: Artificial light gardening. | Indoor gardens.
Classification: LCC SB126 (ebook) | LCC SB126 .H35 2018 (print) | DDC 631.5/83—dc23
LC record available at https://lccn.loc.gov/2018001914

A catalog record for this book is also available from the British Library.

PREFACE

Being obsessed with plants by the time I was 18 years old didn't exactly make me popular at parties. As my original plant friend, Carolyn, will tell you, the two of us could clear a room quickly once the surrounding college partygoers heard us talking about our cactus or cross-pollination. We just couldn't help ourselves, and we're still at it today. However, my status at The University of North Texas as one of only two students (at the time) concentrating in botany, my curious reputation as a gardener, and my college job as a garden-center employee made my phone ring a bit more frequently than my social status warranted. If only I had known I could have made some cash telling anonymous callers how to stop killing their closet so-called tomato plants. But I was never interested in what was actually growing in their closets. I was content to be knee-deep in my outdoor ornamental and vegetable garden, not to mention obsessed with an increasingly large collection of houseplants. But I'm glad those closet gardeners called, because it was the start of my horticultural consulting career—and, as it turns out, this book.

I spent my first two years of university life at UNT as an art major before I switched to biology and botany. As such, aesthetic considerations are infused into all my pursuits, even the scientific ones. The fusion of art and horticulture is natural. Growing plants and food indoors doesn't have to be utilitarian; it can be a beautiful practice that blends into our living space and lifestyles.

Within this book you'll find examples and inspiration for your own attractive plant lighting.

As a graduate student at Michigan State University, my research focused on greenhouse production and flowering of perennial plants. Therefore, you will also encounter some science and math, which may seem a bit confounding at first. If you don't need this information, feel free to skip it. If you have intensive indoor gardening goals, however, the more in-depth how-tos on measuring and calculating your indoor lighting needs will likely form the basis of your long-term success.

As it turns out, writing this book feels like I'm coming full circle to bring the closet garden to light. I hope this book encourages your interest in, and creates new possibilities for, growing plants where you once thought you could not. Perhaps you're on a mission to grow more of your own food or medicinals, even if all you have is a kitchen counter, a guest room corner, or a small closet. You may want to extend your vegetable-gardening season by getting a jump-start on propagation or growing indoors off-season. Having control over your own food source is a powerful feeling. It's good to eat fresh, hyperlocal, and clean. Or maybe, like me all those years ago, you have caught the plant-collecting bug and there just aren't enough windowsills left in your home to feed your growing plant family. In any case, you've come to the right place.

Some plants, such as haworthias, can survive in a windowsill.

INTRODUCTION

My window is bright enough, right? New indoor growers ask this hopeful question all too frequently. While many books and websites insist that you need only a bright window to get your tomato seeds going or your orchids reblooming, this approach can result in a lot of disappointment.

Ambient light levels inside your home are significantly lower in intensity and ultimately different in spectrum than natural outdoor light, especially during the winter months. Even a seemingly bright window may not provide enough light, or the right kind, for young seedlings or fruit-producing plants. Have you ever started seeds indoors in a sunny window, only to watch the tiny seedlings lean so far that they topple over? Perhaps you've tried some beautiful succulents and watched them do the same. Your plants are starving for, and stretching toward, the light.

There are plenty of low-light tropicals and blooming plants that you can grow successfully indoors with good ambient light, and you can maintain certain light-loving succulents for a while in a windowsill. But even a bright windowsill is typically not the right location for plants you intend to harvest for food. The same goes for heavy-blooming plants. Reproduction is an energy-intensive process. Producing flowers, fruit, and seed requires a lot of juice. Plants need enough light, and the right kind of light, to get the job done.

You may have learned that in your outdoor garden, plants that produce large fruit, such as tomatoes, require roughly double the duration and intensity of direct sunlight as plants that produce mostly foliage, such as leafy greens and lettuce. A plant's heritage—that is, the geographical area where it naturally evolved— dictates its requirements for specific light intensity, duration, and spectrum to reproduce successfully. If you are going to invest time, effort, and money in indoor growing, quality supplemental lighting should be your number-one priority.

Before we draw up your shopping list of types of lamps and gear you should be using indoors, or determining how many lamps you will need for different types of plants, we must understand how plants use light; how to distinguish light quality, quantity, and duration; and how to measure each. Artificial lighting is complex,

Many plants will need extra light indoors.

and there is a lot of misinformation in the marketplace, which can ultimately lead to wasted resources. To make good grow-lighting choices and produce better yields, it is necessary to grasp some basic botany and light science.

If you already garden outside, growing plants indoors under lights is a great way to supplement your efforts and extend your seasons and yields. If gardening indoors is your only option, grow lighting can transform your living and eating experiences and bring some much-needed nature into your home.

LIGHT

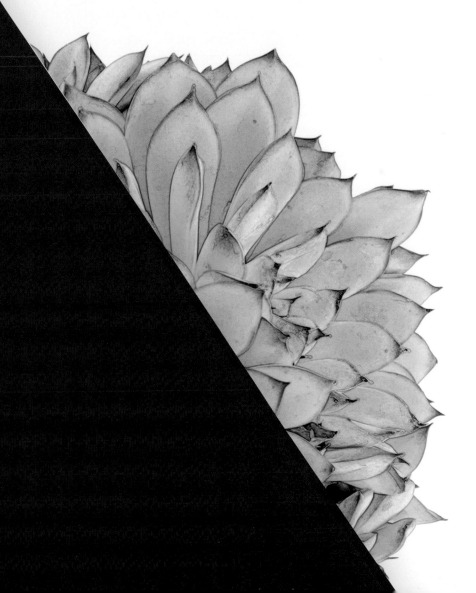

WHY PLANTS NEED LIGHT

Understanding how plants use light is crucial to learning how to grow them successfully, especially indoors. A bit of Botany 101 is a good starting place for all plant enthusiasts.

PHOTOSYNTHESIS

Photosynthesis is a reaction to a transfer of energy. Think of a plant's leaves and green stems as light-energy collectors; nature's very own solar panels. Plant appendages (and several types of microorganisms) use photosynthesis to convert radiant light energy into chemical energy. Light hits the surface of a leaf or a green stem, and specific cells convert the light energy into sugars. These sugars move around the plant, driving various biological functions. A major by-product of photosynthesis is oxygen; hence, we breathe.

When growing plants indoors, your goal should be stimulating and enabling successful photosynthesis. The amount and type of light you provide your plants will ultimately determine their success or failure. Depending on where you live and the time of year, the amount of sunlight penetrating your living room window or filling up your glass greenhouse likely won't be enough to grow many types of plants in an enclosed environment.

Photosynthesis occurs only in the green portions of plants, such as stems and leaves. More specifically, it takes place within the chloroplasts of those plant parts. Chloroplasts are tiny structures inside a plant's stem and leaf cells, like a cell within a cell. Chloroplasts serve as the plant's kitchen and pantry, as they create and store all the needed pigments and food.

Chloroplasts are most likely alien to original plant physiology. Much like our own cellular mitochondria, the organelles we refer to as our cells' powerhouses, chloroplasts were once likely autonomous organisms or bacteria found in the environment—until another organism absorbed them. This organism responded positively to the energy created by its new captive, and the two coevolved. Evolutionary

This philodendron leaf turns light into plant energy.

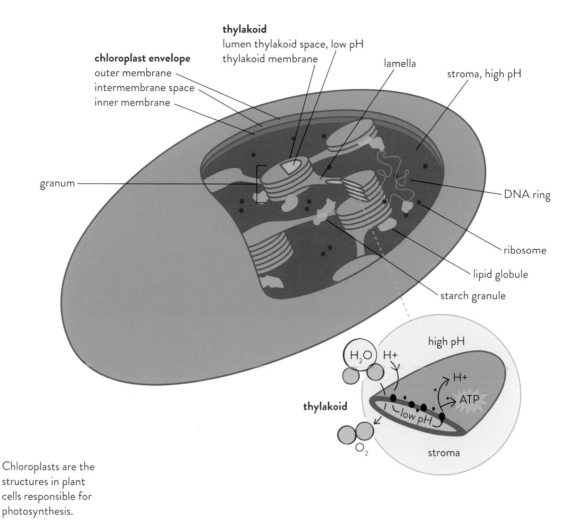

chloroplast envelope
outer membrane
intermembrane space
inner membrane

thylakoid
lumen thylakoid space, low pH
thylakoid membrane

lamella

stroma, high pH

granum

DNA ring

ribosome

lipid globule

starch granule

high pH

H_2O H+

thylakoid

H+

ATP

low pH

O_2

stroma

Chloroplasts are the
structures in plant
cells responsible for
photosynthesis.

theorist Lynn Margulis designed this idea of a beautiful, codependent, mutually
beneficial relationship that is commonly referred to as the endosymbiosis theory.

CHLOROPHYLL

When light reaches a chloroplast, chlorophyll absorbs the pigment inside the chloroplast. What makes plants special is their ability to use this chlorophyll pigment
to convert light into sugars to use as energy. On the light spectrum, chlorophyll

absorbs and employs more red and blue light, leaving more of the green light to bounce back to the human eye. This phenomenon results in the green-colored appearance of most plants.

Two types of chlorophyll are involved in photosynthesis: Chlorophyll A and Chlorophyll B. Chlorophyll A absorbs most of the usable light. Chlorophyll B is a yellow pigment that plays a supporting role by absorbing mostly blue light and transferring it to Chlorophyll A.

Carotenoids, flavonoids, and betalins are additional support pigments—sporting shades of yellow, orange, red, pink, and purple—that also absorb small amounts of light. These pigments are responsible for different colors throughout plant structures. As chlorophyll pigments break down in the autumn months in response to temperature and daylight changes, these carotenoid pigments become visible, resulting in the much-anticipated fall foliage show.

Biological pigments create the rich colors of this begonia foliage.

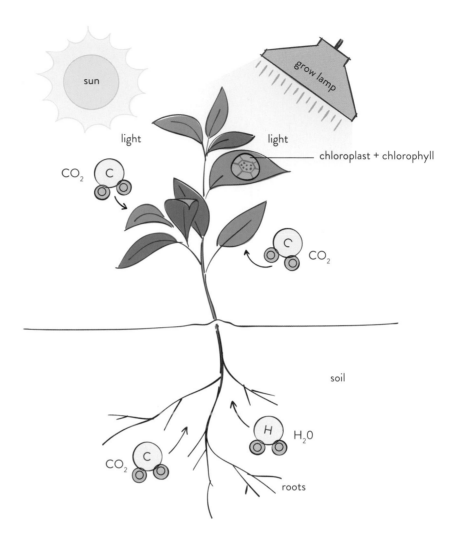

Photosynthesis is the process by which plants use light to generate their own food.

While light is the key trigger and engine of photosynthesis, adequate amounts of water and carbon dioxide are also necessary to the process. For the plant's powerhouses to operate properly, they need carbon dioxide from the air, water from the soil, and light energy from the sun—each in the right amount. If you restrict or eliminate any of these main ingredients, photosynthesis can be impaired or stop altogether.

During the first step of photosynthesis, when light is available, water molecules are split apart. When chlorophyll absorbs light it becomes charged, like a battery, which gives it the ability to split two water molecules ($2H_2O$) into four electrons, four protons, and two oxygen atoms that combine to form O_2, or oxygen gas. This is why we characterize plants as breathing in carbon dioxide and exhaling oxygen. There is a common misperception that plants convert carbon dioxide into breathable oxygen, but the water molecules supply the oxygen gas that plants release back into the atmosphere. The electrons and protons that remain are then stored in proteins within the cell and combined with carbon dioxide in the second step of photosynthesis to form the carbohydrates the plant burns for fuel.

RESPIRATION

While respiration in humans is primarily an exchange of gases in the form of breathing, it is not the same in plants. The primary action of plant respiration is burning sugars generated from photosynthesis to drive growth and development. Basically, respiration is how plants burn calories. While photosynthesis can take place only when there is light, respiration occurs continuously in a plant, just like humans burn calories all the time. Your plants will burn energy even if there is not enough light, water, or carbon dioxide available. If those inputs remain limited or cease, eventually your plant will burn more energy than it can generate. This results in dormancy or death.

This plant ran out of resources and now it's toast.

While plants can generate usable energy within their own bodies, they still rely heavily on many environmental conditions and inputs for photosynthesis to work as intended. When you grow plants indoors in an artificial environment, they become completely dependent on you to create favorable growing conditions and provide them with the right ingredients they need to thrive. As any good baker will tell you, finding the right balance of ingredients and technique takes practice, and every oven functions a bit differently. You will probably flatten a few soufflés—and kill a bunch of plants—before you learn to get it just right.

HOW PLANTS RESPOND TO LIGHT

Understanding how plants see light is the first step in making the right grow-lighting choices. While humans qualify light in terms of visual brightness, plants qualify it in terms of wavelengths, or spectrum.

TYPES OF LIGHT

Not all light is equal. Different types of light both drive and limit photosynthesis, change plant morphology, and influence flowering.

Photosynthetically Active Radiation

When a beam of white light hits a glass prism at an angle, it is then split into different wavelengths of color: violet, blue, green, yellow, orange, and red light. Each of these colors of light measures a different wavelength, falling between 400 nanometers (nm) (violet) to approximately 735 nm (red). This range of visible light is also the range used to fuel photosynthesis. This range of spectrum is known as Photosynthetically Active Radiation (PAR).

In the process of photosynthesis, the red and blue light spectrums most efficiently drive carbohydrate production in plant cells, but all PAR in the 400 to 735 nm range is useful for photosynthesis. PAR is not a measurement of quantity of light, but rather the quality of light. PAR tells you the color spectrum of light it delivers that your plants can use for photosynthesis.

PAR is made up of light particles, or photons, that will eventually strike a leaf's surface. The leaf absorbs these photons in quanta. One photon equals one quantum of light.

You can think of light photons as having calories that plants use for energy, just as our bodies burn the energy delivered from calories in food. Different foods offer different amounts of calories, and so do different wavelengths of light. Light with a short wavelength and high energy (blue) delivers more calories to a plant than light with a longer, lower-energy wavelength (red).

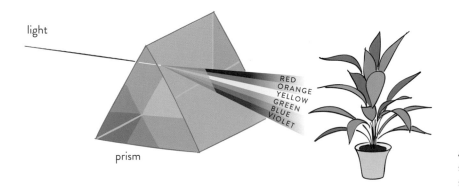

light

prism

RED
ORANGE
YELLOW
GREEN
BLUE
VIOLET

A prism splits full-spectrum light into separate color spectrums.

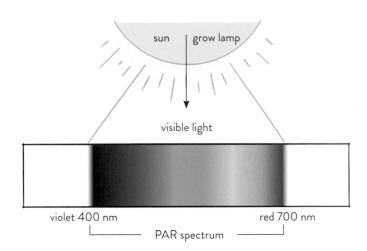

sun | grow lamp

visible light

violet 400 nm   red 700 nm

PAR spectrum

Plants use light that falls into the 400 to 735 nm range for photosynthesis.

Therefore, lamps that emit only blue light, with its shorter high-energy wavelength, are the ticket to tons of giant tomatoes, right? Not so fast. Plants need only a small percentage of their light delivered in the blue spectrum to be able to grow and function well. Red light, despite its longer wavelength and lower energy value, is more efficient at driving photosynthesis than blue light. Cells, structures, and

This light-emitting diode (LED) lamp mixes only blue and red light, resulting in a pink-colored light in the growing area.

pigments (called cryptochromes) that are not involved in photosynthesis also absorb blue light and use it for other operations, such as opening and closing stomata, which are pores in plant leaves that allow for the exchange of water and gases. But only the plant's chloroplasts that contain the chlorophyll, the pigment responsible for driving photosynthesis, absorb the red light photons. While red light may deliver less energy, the plant uses much more of it efficiently to fuel photosynthesis and produce sugars.

The spectrum of light an outdoor garden receives will vary according to geographical location. Sunlight that reaches areas north of the Fortieth Parallel contains more blue light, while sunlight at the equator delivers more red. The atmosphere absorbs light at different levels in each area.

You may have learned that plants don't use any green light, that it all bounces off the plant and makes the plant looks green to the human eye. That is partly true, but it is a myth that plants do not absorb any green light photons or do not use them for photosynthesis. In fact, scientists currently believe that most of the green light spectrum is useful to plants for photosynthesis. It may be that plants have simply figured out a more efficient way to use green light photons, and absorb smaller amounts of it to get the job done. Studies have shown that green light is involved with seedling and leaf development, flower initiation, and plants' use of carbon dioxide and water, and even plays a role in stem growth and height. Green light is not as energy efficient to deliver as red or blue light, but it is easier on the human eye. Green light makes it easier to spot problems such as disease issues or nutrient deficiencies because plants appear their natural color. Green light can also penetrate the leaf canopy more easily, meaning it can reach the lower leaves better than other colors of light. The addition of some green light could help keep the lower leaves on your plants photosynthesizing instead of dying and dropping off.

Ultraviolet

Light that plants can use and humans can see is only a part of all the electromagnetic radiation that surrounds us. Wavelengths that measure below violet light, 100 to 400 nm, are invisible to the human eye and referred to as ultraviolet (UV) radiation. UV radiation does not provide much heat, but it can damage living organisms at the cellular level. The oxygen and ozone in the atmosphere absorbs most UV radiation, which is why ozone is so important.

Plants do not use UV radiation for photosynthesis, and it can cause cellular damage and plant death, but there are some benefits to providing small amounts. Cryptochromes absorb several wavelengths of UV radiation for various beneficial functions. UV damage to plants stimulates them to produce protective antioxidants, resins, oils, and other chemicals that give them flavor. UV light can also toughen up young seedlings so they can more easily transition to higher-intensity lighting without experiencing shock. Some grow lamps include the UV spectrum to boost nutritional value and flavor in edible crops.

If you use grow lamps that emit UV light and you'll be working under or around them, be sure to take the same precautions you would if you were spending time in the sun, such as donning protective clothing and UV-blocking sunglasses.

TOP A tomato plant growing exclusively under yellow-colored light emitted from a high-pressure sodium lamp.

BOTTOM A tomato plant growing under white-colored light emitted from a compact fluorescent lamp.

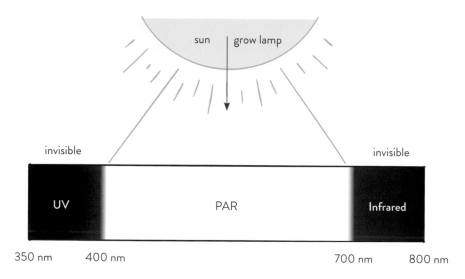

UV and infrared light fall outside the visible PAR spectrum.

Infrared

Radiation wavelengths that fall just above the red light spectrum are called infrared (IR), and this radiation is also not visible to humans. Yet IR radiation makes up about half the solar energy that hits the earth's surface. You can't see IR light, but you feel it as heat. Plants do not use IR radiation for photosynthesis, but it regulates other crucial growth and developmental changes in plants, known as photomorphogenesis. Some lighting systems emit IR at the end of a growing cycle to speed up plant growth and improve blooming.

Plant Biologically Active Radiation

Green plants do not use UV or IR light for photosynthesis, nor do they use radio waves or X-rays, wavelengths even further away on the light spectrum. But these types of nonvisible light do interact with other plant photopigments besides chlorophyll, and they are involved in important biological processes beyond photosynthesis. This wider range of spectrum between 350 and 800 nm is known as Plant Biologically Active Radiation (PBAR), but there are likely biological plant responses within a much larger range of 100 to 1100 nm.

PHOTOMORPHOGENESIS

Overall plant growth and development depends not only on the spectrum of light received, but also on the color combination, color sequence, and duration of each. Plants have evolved to employ finely tuned sensors to use different spectrums of light for different stages of growth and reproduction. Plants use light to regulate developmental stages (such as germination, rooting, stem elongation, leaf unrolling, flowering, and dormant bud development), which is known as photomorphogenesis.

Telling Time

For a plant to grow and bloom on schedule and in the right season, it must be able to tell time. A series of pigments and hormones regulates the time-telling function of photomorphogenesis. Plants produce chemical pigments, called phytochromes, that act as triggers. A blue pigment phytochrome, PR, absorbs and responds to red light. When PR is exposed to red light during the day, it converts into a secondary form of pigment called PFR. When PFR is present in the plant, it tells the plant to produce short, thick stems and determines its overall shape. The presence of PFR is also required to trigger flowering signals. PFR absorbs and responds to IR light over the course of the night. Plants use IR light to tell when it is night, and to determine how much uninterrupted darkness has occurred. In darkness with IR radiation, over time PFR will naturally convert back to PR. This cycle is comparable to the circadian rhythms that help our bodies know when it is time to sleep and rise. The balance of the two forms of phytochrome helps your plant develop properly and on schedule.

Some plant seeds do not germinate until they are exposed to red light. If you have ever tried to grow lettuce from seed, only to be disappointed when the seeds did not sprout, you most likely covered the seeds with soil and blocked them from the light.

Stretching

When plants stretch or grow toward the sun, also known as tropism, they are reaching for more blue light. When sun-loving plants grow in too much shade, their growth slows down and their internode length elongates. The plants stretch to avoid the shade so they can compete with surrounding plants and reach more

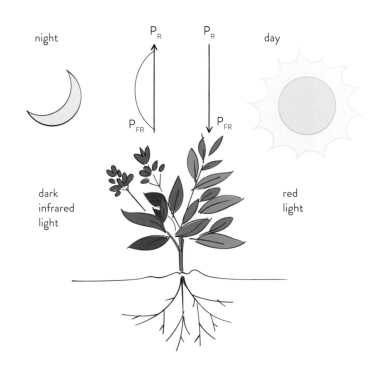

PR to PFR Conversion

night — P_R — P_R — day

P_{FR} — P_{FR}

dark infrared light

red light

PR converts to PFR throughout the day as the plant is exposed to red light. In the dark, when plants can sense only infrared light, PFR converts back to PR.

This arrowhead plant, which sits about 5 feet away from a bright north-facing window, is stretching to reach more light.

RIGHT Some seeds, such as lettuce, require exposure to light to germinate.

light. This internode elongation occurs when the quantity of available light is reduced and the spectrum of available light changes. When blue light is limited or blocked (shade trees block out more blue light than red), it triggers this shade-avoidance stretching.

Red vs. Blue Light

Light-emitting diode (LED) grow-light technology allows you to narrow the spectrum of light you provide to just one color, referred to as nanometer specific light. You can purchase LED grow lamps that emit only a red spectrum of light or a blue spectrum of light (as well as other individual colors, such as orange and green). You can drive photosynthesis efficiently by using only blue and red light together.

While you can use red and blue light exclusively and successfully, only a few species of plants are able to grow well under only red or blue light indefinitely.

Blue light helps control excessive stem elongation, or stretching toward light. It influences how chloroplasts move around in plant cells and helps regulate stomatal opening. Blue light also increases antioxidant levels in crops such as lettuce. You can grow some crops (short-lived ones such as microgreens) using only blue light, which supports ongoing vegetative growth, but don't expect any flowers in certain types of plants. Over the long term without any red light, however, plant leaves may eventually develop too small or become deformed, reducing photosynthesis and other functions.

When you grow plants under only red light, they can put on additional leafy growth, or biomass. This is good for certain crops, such as lettuce. But as plants develop more leaves and other structures, they may not transpire properly or they can stretch, get too tall, or develop oedema (leaf blisters—a common issue in tomatoes grown under only red light) or other problems. If you grow your plants under only red light for too long, chlorophyll production can stop altogether, causing photosynthesis to cease. Plants may even flower too early under only red light in a final effort to reproduce before inevitable death.

You can grow certain quick-turn crops for short periods of time using single-band red or blue LED lamps. For example, if you're growing

ABOVE This LED lamp emits only blue and red nanometer specific light.

BELOW Oedema has become an issue on these pepper plants grown exclusively under high-pressure sodium lamps.

Seedlings growing under full-spectrum fluorescent lamps.

lettuce plants in an enclosed space with artificial lighting, you can start them off using only red light, which causes the leafy greens to grow larger leaves more quickly. However, once the young plants begin to put on more growth, you will need to add 10 to 20 percent blue light to keep them from stretching.

While a little stretching should not concern you, too much can result in weak seedlings that topple over or puny stalks that cannot support flowers or fruits. If your plants are stretching too much, they are not getting enough overall light or they need more blue light.

However, if you want to graft your plants, you may want to encourage them to grow overly elongated stems. Crops such as tomatoes and cucumbers are often grafted onto hardier rootstock, and red and IR light are used to elongate the seedlings to make the grafting process easier. Another scenario for elongated stems is when you are readying a plant, such as cannabis, to flower. If you've been growing it under mostly cool blue light to encourage dense plants, you can then expose it to red light and IR radiation to begin elongation to make room for large flower buds.

Mixing and Matching Light

Switching between color spectrums and types of lamps can help you influence plant-growth habit and trigger different stages of development. You can also mix and match different spectrums of light.

A good technique for most in-home growers is to start seedlings, cuttings, and young transplants using cool full-spectrum LED or cool full-spectrum high-output (HO T5) fluorescent lamps. If you're growing fruiting crops, you can switch plants or lamps to a warmer spectrum (more red light) once you've bumped them up to larger pots to trigger stem elongation and flowering. If you need your plants to flower and they require a specific photoperiod to do so, they will require red light. If you're growing certain plants under only red light, such as lettuce or impatiens, you may find that they won't flower until you add in some blue light.

While this may seem complicated, switching or blending light spectrums is simple, even with basic setups. You can use LED lamps that produce a single spectrum of light, and LED fixtures also come in varying combinations of single spectrums. Some even offer the option within the same fixture to switch from blue only to red only, or to run them together, when you want to shift from vegetative to flowering.

Fluorescent grow lamps are also available in varying spectrums that you can switch out in your fluorescent fixture. You can mix fluorescent tubes of different spectrums in the same fixture to create your own light recipe. High-intensity discharge (HID) lamps are typically offered with specific PAR spectral range outputs, and you can switch plants between these lamps or combine two lamps to achieve your desired light mix.

When I grow tomatoes indoors during winter, I start my seedlings under cool-spectrum HO T5 fluorescent lamps to encourage dense vegetative growth. Once potted up into a larger container, I move the transplants to a grow tent with a ceramic metal halide (CMH) lamp to continue growing them until harvest, or a grow tent with both a warm-spectrum high-pressure sodium (HPS) lamp and a cool-spectrum fluorescent or LED. Or I may grow the young tomato transplants in an open space under a large eight-lamp fluorescent fixture filled with cool-spectrum lamps. As plants grow, I replace half the cool-spectrum lamps with some warm-spectrum fluorescent tubes when I want to trigger more flowering. The switch takes about five minutes. If you use an LED fixture that allows you to use different colors of light within the same fixture, it is as simple as flipping a switch.

TOP Cannabis plants in a vegetative state.

BOTTOM Flower buds on cannabis plant.

Full-Spectrum Light

Full-spectrum grow lamps, whether they skew warm or cool, will emit some percentage of all colors of light within the PAR range. Cool full-spectrum lamps will still emit some red and yellow light, but less of it. Warm full-spectrum lamps

CLOCKWISE
FROM TOP LEFT

This grow tent has two separate lighting sections: one with cool blue light for vegetative plants, the other with warmer red light for flowering.

This grow tent blends blue and red light for growing plants start to finish.

These bulbs are being forced under a mix of red, blue, and white fluorescent lamps.

Lettuce, herbs, and other plants growing together under full-spectrum fluorescent lamps.

will still emit some blue and green light, but less of it. No matter what type of lamp you choose, make sure it provides enough PAR for the type and amount of plants you intend to grow, and remember that you'll get different results from different light colors.

If you're using a complete self-contained in-house growing unit—like one that fits on your kitchen countertop or is built into a cabinet—the type of lighting and spectral shift will be preprogrammed, depending on the type of crop you're growing. All you have to do is stare at your plants as they grow. (And keep them watered, of course.)

PHOTOPERIOD

We all need a good night's sleep. Finding a balance between activity and rest is just as important to plants as it is to us. While there are general guidelines about the amount of sleep we should all strive for, not everyone thrives on the same amount or sequence. Plants are not so different. The length of time you leave your lights on is just as important as the type and intensity of light you provide. Many plants are photoperiodic, meaning they require different durations of light and darkness to shift into different phases of growth or development, such as producing flower buds or forming bulbs.

Home growers can use different regimens to manipulate light and trick plants into flowering and fruiting at abnormal times of the year, or to keep plants in a vegetative state. Learning your plants' photoperiodic needs will help you determine the type of growing setup and lighting cycle you will need to provide.

Most plants require a minimum of 10 hours of light each day for active growth. While your plant might survive with less, it may go into a dormant state if you leave it at that photoperiod for an extended time. It's best to set your lighting timers to a minimum of 10 hours. This is most important if the plants are growing in an enclosed environment without any exposure to ambient light in your home or extra light from a window.

Darkness

Not all plants have a photoperiodic requirement to flower. Geographical origin is the main determining factor. As you get further from the equator, climates grow colder and night periods get longer during the winter. As spring returns and days grow longer (and nights shorter), plants will be triggered to flower when environmental conditions are most favorable and cold temperatures do not threaten reproduction. Closer to the equator, plants evolved to respond to days and nights of almost equal length. Some plants may bloom whenever temperatures are ideal and rainfall is adequate, however, regardless of the duration of light or dark.

Photoperiodically, plants fall into three primary categories: long-day, short-day, and day-neutral. Plants that are not photoperiodic are day-neutral. In other words, long-day plants initiate flower buds when the days grow longer than their

ABOVE Summer-blooming coneflower responds to long days (short nights) to flower.

RIGHT Fall-blooming garden mums respond to short days (long nights) to flower.

critical daylength. Short-day plants will initiate flowering when the days become shorter than their critical daylength. Biologically speaking, however, the opposite is true. It's not the length of daytime, or light, the plants are responding to, but rather the period of uninterrupted darkness. The plant measures the amount of PR phytochrome in its system after a length of darkness. Therefore, long-day plants need short nights to flower, and short-day plants need long nights.

Think about how your plants flower in your garden. Plants that bloom only in summer, such as coneflower, are typically long-day plants; those that bloom only in early spring or fall, such as chrysanthemums, are typically short-day plants. Plants that can flower continuously through the growing season—such as roses, cucumbers, and tomatoes—are day-neutral.

Critical Daylength

When a plant has a photoperiod requirement to flower, it will also have a critical daylength, or period of darkness measured in hours, that triggers it to bloom after a vegetative or dormant phase. The critical daylength is the plant's signal that it is safe to flower. When we expose a 12-hour short-day plant to 16 to 18 hours of daylight or artificial light, followed by 6 to 8 hours of darkness, it will remain vegetative and will not flower. When we expose this same plant to only 12 hours of light and 12 hours of darkness, however, it is triggered to flower. A long-day plant would behave in the opposite manner.

There is another way to trick your plants into flowering more efficiently. If you want to force a long-day plant into flowering faster without lighting it as long, use night-interruption lighting. A few short flashes of a very low level of red light in the middle of the night shorten the dark period, thus initiating the plant to flower without lighting it for a longer period through the day. This can help you save on lighting costs. You can trick short-day plants by delivering a single flash of IR light at the beginning of the plant's dark cycle, when all other lights are off. Biochemically, this adds a couple of hours of calculated darkness by exposing the PFR phytochrome to more infrared radiation in the absence of red light. This approach, with short-day plants, allows you to extend the light period during the day to boost overall growth and yields. You can also reverse both of these methods to delay flowering.

The flowers and bracts of this white poinsettia are initiated by exposure to a specific short-day photoperiod.

Required Photoperiods

Photoperiodic plants fall into two subcategories: qualitative and quantitative. Plants that require a specific critical daylength to flower are classified as qualitative, or obligate. Without that exact amount (or more) of uninterrupted darkness, the plant will never flower.

Poinsettia plants are a classic example of a qualitative (obligate) short-day plant. They will not initiate flower buds until the dark period reaches around 11 hours and 45 minutes, depending on the cultivar. Flower initiation is optimal with a 14-hour period of darkness. Once the plant has reached the critical dark period, it can take anywhere from 8 to 11 weeks at that photoperiod to initiate flowering (which is known as the response time). Temperature also influences the speed of flower initiation during the critical dark period.

Poinsettias are very light sensitive. The 14-hour period of darkness must be uninterrupted, with no exposure to red light, or the plants will never flower or produce the bright red bracts we expect to see around the holidays. Just a single flash of light will restart the time clock and keep the plants in a vegetative state. This is why it's so tough to get your poinsettia to reflower without a controlled growing space and specific lighting.

Depending on your latitude in the Northern Hemisphere, you may be able to encourage a poinsettia to initiate flowering in September and October without special lighting. If your latitude does not produce a long enough dark night period naturally, you will have to provide it artificially. Many commercial growers turn out all those colorful poinsettias by blacking out greenhouses every night for exactly the right amount of time, until the plants initiate flower buds. When I was in graduate school, I spent many a day pulling black cloth by hand to control plant-lighting experiments. These days, most blackout operations are high-tech and automated. You could replicate this process in a sealed grow tent, but I advise buying new poinsettias each year instead.

Manage your light period to keep lettuce and salad greens leafy and to prevent bolting and flowering.

RIGHT Cosmos flowers and leafy lettuce growing side by side under the same lighting photoperiod.

Getting cannabis to produce flower buds requires a strictly controlled lighting regimen.

Beneficial Photoperiods

Plants that do not require an exact photoperiod before they are triggered to flower, but will bloom faster or better with a specific photoperiod, are classified as quantitative, or facultative. Lettuce is a classic quantitative long-day plant. While you might start your tiny lettuce seedlings with a long photoperiod of 16 hours, if you continue growing them indoors you must reduce that lighting period to 12 to 13 hours after you transplant the seedlings into their final container. If you continue growing them with 16 hours of light, the plants may bolt and go to flower too early, ending your harvest.

A quantitative short-day plant may bloom earlier or better if it has longer nights. A quantitative long-day plant may bloom earlier or better when given shorter nights. Both plants will eventually flower, however, even if the photoperiod is not ideal. If you're growing different crops together, be sure to take into account their photoperiodic needs. Cosmos is a quantitative short-day plant. You can grow cosmos indoors with long daylengths, and after several months they will eventually flower. But if you grow them with short days and long nights, they will start flowering in just a few weeks. I like to grow cosmos indoors as a cut flower, so I plant it alongside my lettuce. By growing these two crops together with shorter daylengths—12 to 13 hours, versus the 14 to 16 hours you may use for tomatoes and peppers—I'm slowing down flowering in my quantitative long-day lettuce while speeding up flowering in my quantitative short-day cosmos.

Marijuana, *Cannabis sativa*, is a qualitative short-day plant. To remain in a vegetative state, plants require 16 to 18 hours of continuous light followed by 6 to 8 hours of uninterrupted darkness. Once you are ready to force them to flower, reduce their daylength to 12 hours, followed by 12 hours of uninterrupted darkness, for about two weeks.

Photoperiod also influences tuber and bulb formation in crops such as garlic, and there are other, more complicated photoperiodic requirements in plants, but understanding the basic classifications is sufficient for home growers.

You might wonder if keeping light on your plants 24 hours a day will help them grow faster or bloom better. While you can light young seedlings for 24 hours, as well as some quick-turn crops such as microgreens or whole heads of lettuce, long-term 24-hour lighting can be detrimental to many plants, or it is wasteful because certain plants won't produce any better with 24-hour lighting than they will under less lighting. If your plant has a photoperiod requirement, it won't flower under 24 hours of continuous light.

If you do not take the time to research your plant's photoperiodic needs, you might be left wondering why it never flowers—or why it keeps flowering when you don't want it to.

You can use simple lamp timers or more sophisticated digital timers to turn your grow lamps on and off according to a set lighting schedule.

Examples of photoperiodic plants

PHOTOPERIOD	EDIBLE	ORNAMENTAL
qualitative (obligate) long-day	chicory, cilantro, dill, endive, oregano, spinach	bachelor's buttons, fuchsia, gazania, lobelia, monkey flower, sweet pea, strawflower, certain hybrid petunias, such as 'Purple Wave'
quantitative (facultative) long-day	beets, carrots, chard, lettuce, mint, peas, thyme	ageratum, calendula, dianthus, pansy, grandiflora petunia, snapdragon, salvia, sunflower, viola
qualitative (obligate) short-day	cannabis, common bean, potatoes, sweet corn, sweet potatoes, strawberries; onions and garlic for bulbing	African marigold, fuchsia, hyacinth bean vine, poinsettia
quantitative (facultative) short-day	potatoes, sweet corn, sweet potatoes, yams for root development	cosmos, globe amaranth, moonflower, morning glory, zinnia
day-neutral	eggplant, peppers, tomatoes	amaranthus, centranthus, cleome, stock, verbascum

*Ornamental variety information referenced from R.M. Warner.

MEASURING LIGHT

Having the right spectrum of light is not enough to grow healthy plants. You also need the right quantity to fuel photosynthesis. While accurately measuring light for plant growth might seem complicated—and the fine details are beyond the scope of this book—an introductory understanding of how to measure and manipulate light will help take your indoor-growing skills to an entirely new level.

Many indoor gardeners start out germinating seeds with a basic kit-type plant-lighting system or with only a few small indoor houseplants. If that's the case, you probably don't need to pull out a light meter or a calculator, and you can skip much of this chapter. If, however, you want to produce good harvests of fruits such as strawberries and tomatoes, or plants that have tricky flowering requirements such as cannabis, then let's crunch some numbers—and brain cells.

COMMON MEASUREMENTS OF BRIGHTNESS

When you go online or to the hardware store to buy lamps, manufacturers usually provide a few light measurements. The most common are watts used and lumens delivered by the lamp, along with Kelvin. You may also find references to lux or footcandles. While these measurements are relevant to how much energy your lamp uses and how bright it will make your kitchen, they aren't necessarily relevant to plant growth. Nevertheless, you need to understand these metrics because they may be the only information you receive.

Lumens

The measurement of lumens tells you how much light you can expect to see from a given lamp, or the total amount of light output. Lumens per watt (LPW) is used to measure the efficiency of a lamp, or how much electricity a lamp converts into light versus heat. Remember, plants and people see and use light differently. A light source may seem bright to your eye, but that doesn't mean it's better for plants. Humans perceive yellow and green light to be brightest. A lamp that emits more yellow and green light will have a higher lumens or lux rating because it looks brighter, but it won't necessarily be better for photosynthesis.

While lumens or LPW measurements don't tell you about the type of light your lamp delivers, they can help you determine its efficiency. You don't want your lights producing a lot of unwanted heat (versus usable light), which can damage your plants and make it harder to control your environment, so the better the LPW measurement, the better your results.

Lux

Lux is the intensity of the lumens over a defined area in square feet. Lux is important in terms of calculating how many lamps are adequate for a particular space given the light-output capacity of the lamp. Lux measurements can tell you how evenly your light is distributed over the area. When you are deciding how bright your new kitchen light fixtures should be, lumens and lux are your standard metrics. Lamp packaging may include measurements of lux at varying distances to surfaces.

Footcandles

Footcandle measurements are another traditional method of determining visual brightness for outdoor sunlight and indoor lighting. Footcandles don't tell you directly how much or what type of light your lamp is delivering. Instead, you can use footcandles to get a basic idea of whether your lamp is providing a lighting environment similar to full sun, partial sun, or shadier conditions. Direct sunlight outdoors measures about 10,000 footcandles (fc), whereas a cloudy day may deliver only 100 fc. Areas under shade trees typically measure 50 to 100 fc. Full-sun plants, such as tomatoes and roses, need light in the 7000 to 10,000 fc range, while shade-loving ferns can survive on 100 fc. One footcandle equals one lumen per square foot, so 1 footcandle equals approximately 10 lux.

TOP You can start fruiting plants, such as these alpine strawberries, from seed and grow them under bright lights indoors.

BOTTOM Most grow lamps will provide you with a lumens output, which is a measure of visual brightness.

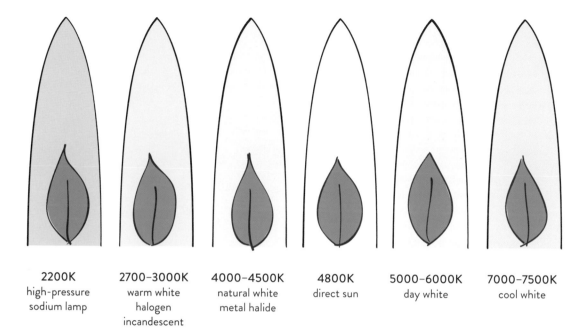

2200K	2700–3000K	4000–4500K	4800K	5000–6000K	7000–7500K
high-pressure sodium lamp	warm white halogen incandescent	natural white metal halide	direct sun	day white	cool white

Kelvin ratings provide a visual temperature reference for different types of lamps.

Kelvin

Kelvin temperature is used to describe the visual color of light that a lamp emits. Technically, it is a measure of how white a piece of tungsten steel is when it's heated to a specific temperature in degrees Kelvin. A lamp with a Kelvin measurement of more than 5000 Kelvin (K) produces more light in the cool blue spectrum. Lamps with a lower measurement, in the 3000K or lower range, produce more light in the warm red spectrum. Kelvin temperature is not a measurement of actual hotness or coolness of the lamp, but rather the visual temperature of light, which can have a big impact on how your indoor space looks. Cool fluorescent lights over the bathroom mirror? No one wants to see that first thing in the morning.

Much like lumen and lux or footcandle measurements, a Kelvin measurement does not give you specifics on the actual light spectrum delivered to your plants. It is a visual approximation of how much blue or red light your plant may receive from a given lamp. For example, a T5 fluorescent lamp with a 6500K rating will skew to the cool side, meaning it provides more light in the blue spectrum. But it may be labeled a full-spectrum lamp because it will also deliver a small amount of red light, along with some orange, yellow, and green light. From a measurement of 6500K, you could infer that this lamp is most appropriate for growing plants in their vegetative state, such as seedlings, salad greens, and leafy herbs. A HPS lamp with a Kelvin rating in the 2700K range will skew to the warm side, meaning it provides more light in the red spectrum. Again, as a full-spectrum lamp it will still emit small percentages of cool-spectrum colors. A warm-spectrum light will be more useful for plants that need to flower and fruit.

Watts

Watts tell you how much energy your lamp is using and how expensive it may be for you to run the lamp. You can use watts to estimate the size or quantity of lights you may need. But just like a Kelvin rating, watts do not tell you how well a lamp will grow your plants. Light, not electricity, grows your plants.

While a lamp can seem very powerful because it uses many watts, if it is not efficient, it may convert too much of that electricity into unwanted heat instead of light. Also, a lamp's advertised wattage may not match the amount of electricity it draws. For example, a 3-watt LED lamp may only pull 2 watts of electricity when turned on. Thus, it will be less powerful than you expect, and it will deliver less usable light. If you base your lighting choices purely on wattage, you may end up with an unnecessarily large electricity bill and a lot of wasted energy in the form of unwanted heat.

Pay attention to the volts and amps ratings for your lamps. If you purchase a lamp that has a different voltage than your outlets, you'll need special circuits and wiring. Learn the amp capacity of your fuse box and breakers. If the collective amp load of your grow lamps will push you over 75 percent usage on a given circuit, you may need to have an electrician add a separate circuit for your grow lamps.

HOW TO TAKE ACCURATE MEASUREMENTS

There are a few factors to consider if you want to accurately measure the quality and quantity of light your lamp provides specifically for plant growth. Before you pull out your calculator, remember there are inputs and outputs. Inputs include

These fluorescent grow lamps have a 6500K rating, which means they emit a cool-colored full-spectrum light.

LEFT This compact fluorescent bulb uses 26 watts to operate, and it will fit into many home fixtures.

electricity usage, measured in watts. Outputs are the amount of light and heat produced. Your plants respond only to the outputs they can use biologically, so these are the real metrics to define. Once you know the PAR spectrum of the lamp, you need to determine how much overall light it puts out, how much of that light is available to your plants, and the cumulative amount of light the plant will receive during its photoperiod or duration of lighting. These three output measurements are Photosynthetic Photo Flux (PPF), Photosynthetic Photon Flux Density (PPFD), and Daily Light Integral (DLI).

If you can get a handle on these three key measurements, you will be a pro at plant lighting. Good lighting manufacturers generally include this information with their lamps or on their websites.

PPF

PPF is the measurement of the PAR your lamp emits per second. This amount of light is measured in micromoles per second, expressed as μmol/second (a micromole is one millionth of a mole). A mol is the unit of measure for photons. PPF can be tricky to determine, however, because until recently most lamp manufacturers did not measure it or list it on their products.

Lamp manufacturers may refer to the energy efficiency of a lamp in terms of the units of PPF generated per unit of power used, expressed as μmol/s/Watt. This measurement is known a PPF efficacy. You may also see PPF efficacy expressed as μmol/J, where J refers to joule, a derived unit of energy. Efficient lamps can generate 1 μmol/s/W, or 1 μmol/J, or more. Lamps that emit more red light, such as HPS lamps, tend to put out slightly higher efficiency values, given that red light is more efficient for photosynthesis.

Check the manufacturer's packaging for the PPF and PPF/W values, or contact them directly for this information if it is not provided with the lamp.

PPFD

Next, you will need to determine how many PAR photons produced every second from your lamp land directly on your plant leaves. PPFD is measured over one square meter and in micromoles per square meter per second ($\mu mol/m^2/s$) at a specific distance from your plants. It is the measurement of the density of useful light your lamp provides over a specific area.

For example, a cloud (lamp) produces a certain number of raindrops (light photons) every second. PPF measures that number. If the cloud is large, the raindrops it produces are spread out and there is a lower density of raindrops in a given area. If a much smaller cloud produces the same number of raindrops per second in the same area, the drops (photons) will be much denser in their defined delivery area, and you will accumulate more total rainfall (light photons) more quickly in that space.

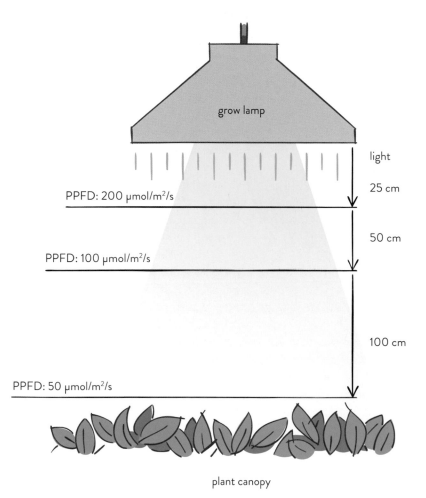

grow lamp

light

25 cm

PPFD: 200 μmol/m²/s

50 cm

PPFD: 100 μmol/m²/s

100 cm

PPFD: 50 μmol/m²/s

plant canopy

PPFD measurements tell you how much PAR is delivered at specific distances from the lamp, based on the lamp's PPF output.

Good plant-lighting manufacturers will not only provide the PAR and PPF measurements of their lamps, but also the PPFD value measured at a specific distance and area underneath the lamp fixture. This information will tell you how much useful light your plant will receive when placed at that distance under your lamp and over a specific area. Some horticultural-lighting company websites even provide calculators that you can use to determine your PPFD and lighting needs before you buy, but these measurements are often available only for high-intensity grow lamps.

There is a lot of variation in how manufacturers measure light quality and the outputs of their lamps, as well as differences in how they determine the

CLOCKWISE
FROM TOP LEFT

PAR calibrated meters are available with attached and remote sensors so you can measure light in different locations and over different lengths of time.

I use my quantum flux meter to decide which lamps I will use to grow what plants at which stages of development. And to impress people at parties.

Using my quantum flux meter, I can measure PPF and PPFD output from my fluorescent grow lamps. Not enough light for the full-sun strawberry plant on the shelf, but just enough for small light-sensitive African violet plantlets I propagated.

If you thought that window was bright enough, think again. This poor begonia won't last long. There is barely enough light for most plants to survive long term.

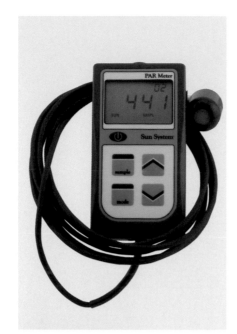

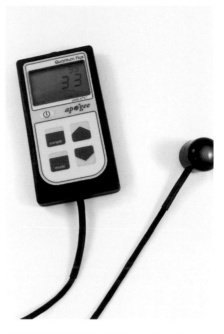

measurements they provide. Some incorrectly use the terms *PPF* and *PPFD* interchangeably as the same measurement. You can take the science into your own hands and determine accurate measurements with a PAR quantum flux meter (not to be confused with a flux capacitor).

Don't confuse a quantum flux meter with standard visual light meters or Watt meters. Using a Watt meter and a standard light meter that measures lumens, you can calculate LPW outputs for a given lamp, which helps you assess efficiency. You can also use standard light meters to measure footcandles of light. But remember that standard light meters, which measure in footcandles or lux, are not accurate for determining plant light needs.

To measure light for growing accurately, buy a calibrated PAR quantum flux meter, which is a light sensor connected to a light meter. Botanists and horticulturalists use these tools, which measure the spectral region between 400 and 700 nm, to obtain the incident PPF your lamp delivers each second. Newer meters are specially calibrated for the wider nanometer range commonly emitted by LEDs. Good quantum meters are not cheap, but if you're interested in high yields they are an excellent investment.

You can use your quantum flux meter to take PPF measurements to calculate PPFD. To determine your average PPFD for a given lamp and space, take several measurements around the actual plant canopy (only the area where plants will be growing) from a specific distance from the light fixture. Average these numbers to get a meaningful PPFD value. This is necessary because there are differences in light delivery with different-size fixtures, and the PPFD degrades when the lamp is further away from your plants.

CALCULATOR ALERT

PPF output average (taken from several locations around your plant canopy using your quantum flux meter): 1000 µmol/s

Growing Area: 5 ft. × 5 ft. = 25 ft^2 = 2.32 m^2

Degradation: We'll use a common estimate of 20 percent loss of light that may not reach all of your plant canopy, which translates to 80 percent efficiency at the given distance. This is sometimes referred to as the coefficient of utilization (CU).

This equals 1000 µmol/s × 0.8 = 800 µmol/s

PPFD = 800 µmol/s ÷ 2.32m^2 = 344 µmol/m^2/s

A full-spectrum sensor measuring PAR from blue and red LED lamps.

Now that you have all the tools and equations to calculate how much useful light your lamps are going to produce, you can predict your yields with frightening accuracy, right? Hold on a second. Just as manufacturers may use different methodologies to calculate their lamp ratings, the PPF or PPFD measurements taken with flux quantum meters are not perfect. They can vary depending on the type of lamp from which you are taking measurements (MH versus LED) and the spectrum output of the lamp. (Remember, red light is more efficient at stimulating photosynthesis than blue light.)

The difference in the efficacy of the varied color spectrums of light on actual photosynthesis is called the quantum efficiency curve, or yield photon flux (YPF) curve. If you take measurements using a standard quantum flux meter under a lamp that generates only blue light, you are going to get good PPF and PPFD values, even though blue light is less efficient at stimulating photosynthesis than red light. Remember to account for the spectrum of light your lamp provides and how plants use each spectrum. There are now more advanced full-spectrum quantum flux sensors that are better calibrated for measuring PPF for LED lamps.

If you don't want to spend the money on a quantum flux meter but you do want a very basic estimate of your light levels, you can use a simple sunlight meter designed to rate outdoor garden light. These basic light meters help you identify where you have enough light for plants that need sunny or shaded locations, with measurements delivered in footcandles or rated full sun, partial sun, or full shade.

DLI

The third measurement, and potentially the most important for your plants, is Daily Light Integral (DLI). This is a cumulative measurement of the total number of PAR that reaches your plant during the provided photoperiod, the amount of time your plants are lighted. DLI is measured in moles per square meter per day, or $mol/m^2/d$. You use your PPFD measurement to calculate your DLI.

DLI is important because the amount of photosynthesis that can occur using the light you provide ultimately depends on how many hours a day the plant receives that quality, quantity, and density of light. The longer you light your plants, the more individual light photons they will absorb, thus generating more photosynthesis. More photosynthesis means more tomatoes.

DLI, in conjunction with light spectrum, determines overall yields. Good yields are the primary goal when you are growing edible and flowering plants. In greenhouse and hydroponic food-production growing systems, DLI is the most important measurement, and professional growers frequently start with this number.

Natural light levels and DLI outdoors change dramatically through the seasons. In winter months, ambient DLI is typically too low to get good yields on many crops, whether outdoors or in a glass greenhouse. Supplemental artificial lighting is the only way to deliver adequate DLI for producing successful crops. DLI inside your home, in front of your windows, also drops dramatically in winter. While your favorite plants might get enough light in a northern-exposure window in summer, they may not under the short, dark days of winter.

To produce a good harvest, this tomato plant needs full-sun light levels delivered for a long enough duration.

Lamps that allow you to raise and lower them on the same fixture are handy for growing plants with different light requirements.

ABOVE RIGHT These cucumber plants were started from seed at the same time. The container on top was grown under a CMH lamp with a higher PPFD, while the container on the bottom was grown under a CFL with a lower PPFD. The plants grown under the higher-intensity lamp are far ahead of those grown under the lower-intensity lamp.

Different plants have different requirements for DLI to photosynthesize properly. As a very general rule you can group them into basic categories:

1. Full-sun very high–light plants, such as tomatoes, need 18 to 30 mol/m²/d.
2. Part-sun/part-shade medium-light plants, such as lettuce, need 12 to 16 mol/m²/d.
3. Low-light plants, and vegetative seedlings and cuttings of many edible crops, typically do well with 6 to 10 mol/m²/d.
4. Heavy-shade plants do well with 3 to 6 mol/m²/d.

Typically, the higher the PPFD delivery of a grow lamp, the more expensive it is. You can run a less expensive lamp with a lower PPFD value for a longer photoperiod and rack up the same DLI that your plants need. Just be sure not to run the lamp longer than the plants can tolerate. You can also choose a more expensive lamp with a higher PPFD and run it fewer hours to deliver the same DLI. Provide a long enough photoperiod if one is required.

Another trick is to raise or lower a grow lamp to deliver higher or lower intensities of light to your plant. Lowering the lamp reduces the light footprint but increases the density of light delivered. Raising it increases the light footprint but

CALCULATE THE DLI YOUR LAMP DELIVERS

Calculator alert: PPFD × number of hours of light on per day × 0.0036 = DLI

Here, 0.0036 is the number of seconds in an hour, divided by one million. Convert the number of hours of light into a decimal. For example, 13½ hours becomes 13.5, and 11 hours and 45 minutes is converted to 11.75.

200 µmol/m²/s × 12 h × 0.0036 = 8.64 mol/m²/d

You would have to run a lamp that has a PPFD of 200 µmol/m²/s for 12 hours to generate a DLI of 8.64 mol/m²/d. That lamp, run for 12 hours a day, would then be good for low-light shade-loving plants. Running the lamp for 12 hours a day, however, would not provide an adequate DLI for tomatoes.

Now, let's take that same lamp with the same output and run it for a few more hours a day:

200 µmol/m²/s × 18.5 h × 0.0036 = 13.32 mol/m²/d

You still don't have a DLI that's adequate for tomatoes, but now you can successfully grow lettuce, leafy greens, herbs, and other medium-light plants.

You might be wondering if you could grow tomatoes, which need a DLI between 20 and 30 mol/m²/d, if you ran this same lamp for 24 hours a day. Unfortunately, tomato production can decline with 24 hours of light, so that's a no-go. Your solution is a more powerful lamp with a higher PPFD.

400 µmol/m²/s × 18.5 h × 0.0036 = 26.64 mol/m²/d

Now we're talking tomatoes. You'll need a lamp with double the PPFD levels that you would require for lettuce to grow healthy tomatoes within their ideal photoperiod range (14 to 20 hours of light). If you want to run your lamp for a shorter time, you'll need a lamp with an even higher PPFD.

You can mix and match different PPFD levels and daylengths and use the same equation to calculate your best combination of lamps and lighting durations.

lowers the light density. This is how you can use the same lamp to grow a full-sun plant and a shade-loving plant.

PBAR Flux

Plant growers have started to emphasize PBAR, recognizing that these additional invisible wavelengths are key components to good indoor plant lighting. Keep an eye out for newer PBAR flux numbers on grow lamps, which combine the PPF and the additional PBAR outputs for a total measurement of µmol/s. If your lamp comes with a PBAR flux measurement, you may also get a PBAR efficacy measurement.

CONVERTING LUMENS

What if you don't have access to PPF or PPFD measurements for your lamp manufacturer, don't want to buy a PAR meter, or simply want to rely on the more traditional measurements? What if the lamp manufacturer has provided the wrong light measurements with your lamp?

As the label below shows, the lamp manufacturer included measurements of lux, but no PPF information. In fact, the label on the right mixes up PAR and lux measurements in an attempt to provide PPFD measurements at different distances from the plant. But, as you now know, lux and PAR are distinct measurements. Unfortunately, far too many lamps include this kind of incorrect information.

Amateur indoor-growing enthusiasts and professional greenhouse growers frequently use metrics such as lumens, lux, Kelvin, and watts to calculate indoor plant-lighting needs and lamp outputs because of a simple lack of relevant ratings from lamp manufacturers and an evolving understanding of light science. Conversions were developed to help growers make useful comparisons, for specific types of grow lamps, between lux and units that are meaningful to plant growth. These conversions are not exact, so there will be some degree of error in the results, but you can use them to make an estimated comparison between lux and PPF or PPFD output. According to researchers Richard Thimijan and Royal Heins, who helped develop the conversions, the calculation error can be about 10 percent, depending

The label from an LED grow lamp I purchased.

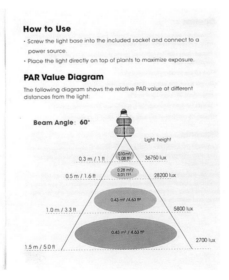

Specifications

Model	TT-GL23
LED Color	Red & Blue
LED Ratio	Red (660nm 3pcs; 630nm 7pcs) Blue (460nm 2pcs)
Power	36W
Socket Model	E27
Input Voltage	AC 85-264V
Input Current	650mA
Frequency	50 / 60Hz
Lux Value	1 ft / 36750 lux 1.6 ft / 28200 lux 3.3 ft / 5800 lux 4.9 ft / 2700 lux
Suggested Height	1.7 ft – 5 ft
Illuminated Area	0.60 – 5.46 ft²
Lifespan	Up to 50,000 hours

Uses

· Ideal for horticulture, indoor gardening, plant propagation, and food production, including indoor hydroponics and aquatic plants.

· Suitable for growing houseplants, such as orchids, roses, peppers, tomatoes, basil, lettuce, herbs, kale, spinach, wheatgrass, broccoli, wildflowers, cucumbers, and many other fruits and vegetables.

How to Use

· Screw the light base into the included socket and connect to a power source.

· Place the light directly on top of plants to maximize exposure.

PAR Value Diagram

The following diagram shows the relative PAR value at different distances from the light:

Beam Angle: 60°

Light height

0.3 m / 1 ft   0.10 m² / 1.08 ft²   36750 lux

0.5 m / 1.6 ft   0.28 m² / 3.01 ft²   28200 lux

1.0 m / 3.3 ft   0.43 m² / 4.63 ft²   5800 lux

1.5 m / 5.0 ft   0.43 m² / 4.63 ft²   2700 lux

PPFD to Lux Reference

PPFD (μMOL M^{-2} S^{-1})	LUX (SUNLIGHT)	LUX (HPS)	LUX (METAL HALIDE)	LUX (FLUORESCENT)
10	540	820	710	740
10	540	820	710	740
100	5400	8200	7100	7400
200	10,800	16,400	14,200	14,800
300	16,200	24,600	21,300	22,200
600	32,400	49,200	42,600	44,400
1000	54,000	82,000	71,000	74,000
2000	108,000	164,000	142,000	148,000

Provided by Apogee Instruments

on differences in lamp manufacturing, lamp age, and how you operate the lamp. These conversions can help you make quick basic determinations about buying the right type of grow lamp for the size of the space and type of plants you intend to grow.

LEDs are not included in the conversion tables because there is a wide spectral output difference between all the various types. Standard conversions do not apply for LEDs. To take accurate PAR measurements from your LEDs, you will need to rely on a full-spectrum PAR quantum flux meter or spectroradiometer.

Conversion Factors: The Conversion from PPFD (μmol m⁻² s⁻¹) to Lux Varies Under Different Light Sources

PPFD (μMOL M⁻² S⁻¹) TO LUX		LUX TO PPFD (μMOL M⁻² S⁻¹)	
sunlight	54	sunlight	0.0185
cool white fluorescent lamps	74	cool white fluorescent lamps	0.0135
mogul base high-pressure sodium lamps	82	mogul base high-pressure sodium lamps	0.0122
dual-ended high-pressure sodium: ePapillion 1000W	77	dual-ended high-pressure sodium: ePapillion 1000W	0.0130
metal halide	71	metal halide	0.0141
ceramic metal halide (CMH942): standard 4200K color temperature	65	ceramic metal halide (CMH942): standard 4200K color temperature	0.0154
ceramic metal halide (CMH930-Agro): 3100K color temperature, spectrum shifted to red wavelengths	59	ceramic metal halide (CMH930-Agro): 3100K color temperature, spectrum shifted to red wavelengths	0.0170

Multiply the PPFD by the conversion factor to get lux. For example, full sunlight is 2000 μmol m⁻² s⁻¹ or 108,000 lux (2000 × 54).

Multiply the lux by the conversion factor to get PPFD. For example, full sunlight is 108,000 lux or 2000 μmol m⁻² s⁻¹ (108,000 × 0.0185).

Provided by Apogee Instruments

CONVERSION EXAMPLE

Your Space: 121.92 cm × 121.92 cm (4 ft. × 4 ft.) = 1.48 m²

Your Lamp: 250W CFL (11,000 lumens according to lamp package)

lux = (total lumens) ÷ (total area in square meters)

= 11.000 ÷ 1.48

= 7432.43

PPFD = lux × conversion factor listed in table above for a fluorescent lamp

= 7432.43 × 0.0135

= 100.33 PPFD

If we kick back to our DLI equation:

100.33 × 12 hours × 0.0036 = 4.33 mol/m²/d

If you run this CFL for 12 hours in this area with no other ambient light, it will provide enough light only for heavy shade–tolerant plants. You could run it for 18 hours and have enough light for certain low-light plants, seedlings, and vegetative propagation. But if you plan to grow tomatoes, you'll need to buy a lamp with at least three times the lumens output for your space.

GROW LAMPS

Now that you are armed with an in-depth understanding of how your plants respond to light and how much light they need, you can choose the right lamps for your home-growing needs. Many indoor gardeners wonder if they can use shop lights. It's possible, but you will not get the same results, efficiency, or aesthetics from standard shop lights or other common household lamps. You will likely end up with a lot of unwanted heat, wasted electricity, and less useful light. As a general rule, the less expensive the purchase price of the light setup, the more expensive and less effective it is to run. You will get much better results if you provide the right intensity of selective or full-spectrum light that is specific to plant growth.

Here are a few key factors you should consider when choosing the type of lamp, or combination of lamps, you will use to grow your plants: the spectrum of light provided; that is, the balance of cool and warm colors and how much of each is provided; the usable quantity of light the lamp provides; the amount of heat output from the lamp; and the amount of light, or the number of hours, you intend to light your plants given their daylength requirements and the lamp's energy efficiency.

Don't base your lighting decision solely on the number of electrical watts the lamp needs. Watts are an input requirement of the lamp, not an output of resources for your plant.

When comparing grow lamps, don't rely solely on the listed lumens of the lamp. Lumens are a general measure of brightness for the human eye, not a measure of usable light for photosynthesis.

Ultimately, the type of lamp you will need depends on the types of plants you are growing (leafy, flowering, or fruiting), the number of plants you will be growing, the size of the plants, and the dimensions of your growing space.

If you simply want to keep a single houseplant or a blooming orchid in your office or living room, you need only a single smaller spotlight fixture. A small crop of leafy greens, microgreens, or herbs is also simple to light with a small-footprint

Spotlight LED grow lamps are ideal for lighting individual houseplants.

BELOW LEFT This self-contained LED growing unit provides temperature and humidity control.

BELOW Countertop growing systems allow you to tuck small greens and herbs into your kitchen or office.

A traditional incandescent bulb can be bright, but hot.

fluorescent or LED setup. Many self-contained home-growing units are already complete with full-spectrum LED lamps for growing seedlings and microgreens. If, on the other hand, you intend to grow groups of fruiting crops such as tomatoes, citrus, or cannabis, you will have to upgrade the type and quantity of your grow lights.

Grow lamps will emit decreasing amounts of light over their life span. This is known as depreciation. The packaging or product description usually includes an approximate number of hours you can expect to receive full PAR. If the life of the lamp is 10,000 hours, the PAR may begin to degrade at about 8000 hours. The number of hours a day and number of days per year you're running the lamp can help you determine when to replace it.

LOWER-INTENSITY LIGHTING

There is a wide range of grow lamps available to home growers. Let's start with the lowest-intensity lighting choices, and then we'll work our way up to more light-intense options.

Incandescent Lamps

Incandescent lamps are frequently the first choice of new indoor gardeners, as well as those who want to grow an individual plant or a small collection of plants. While you can keep standard foliage and some easy-care bloomers growing with an incandescent grow lamp, it is the least efficient and effective option.

An incandescent lamp converts most of the electricity it uses to heat, leaving only a small percentage of the remaining energy to deliver usable light to your plants. If you have ever touched an incandescent bulb and burned your fingers, you know how hot it can get. The low output of usable light and high output of unwanted heat makes an incandescent lamp a less-than-desirable choice for growing any sort of edible crop, especially those that produce fruit. Plus, incandescent bulbs are typically designed for individual spotlight-type fixtures, so you would have to set up quite a few to light a row of seedling trays or larger potted plants.

An incandescent grow lamp can work in a darker space where you may need more ambient lighting for an individual foliage or blooming plant. It can be useful to site an incandescent grow lamp in a fixture that also serves as a directional reading lamp. Remember, however, that your plant will need that lamp to remain on for much longer than you may need it for reading or lighting the room. This will use more electricity and generate much more heat than you may want. Incandescent bulbs typically have a life span of less than 1000 hours, so be prepared to replace them frequently.

Fluorescent Lamps

Fluorescent lamps are popular with home gardeners for grow lighting. Fluorescent tube lamps are used for all sorts of purposes, including standard shop light and office light. They provide two to three times more total light than incandescent lamps, using the same amount of electricity while producing less heat, so they are a much more effective and energy-efficient option than incandescent lamps. They are also available in different diameters and lengths. You can expect a fluorescent tube to give you about 20,000 hours of light for growing. That's roughly three years if you're using the lamp to light plants for 16 hours a day every day of the year. Be aware, however, that efficiency can degrade over the age of the lamp. You can use fluorescent tube lamps in just about any fixture that holds traditional fluorescent tubes, so they are an easy and affordable solution for most home growers.

Fluorescents are typically labeled T15, T8, or T12. The T designates size by diameter. T5 tubes are about the diameter of a dime, while a T8 is closer to a nickel. T12 tubes are the largest. As their diameter grows smaller, fluorescent lamps get more efficient. A T8 fluorescent tube will be more efficient than a T12, and the T5 tubes are more efficient than both.

If you want to get the most growing bang for your buck with fluorescents, look for those labeled as high output (HO) or very high output (VHO). HO T5 tubes

You can use incandescent bulbs in household lamp fixtures to light individual plants, but you will need to leave them on most of the day.

These T5 fluorescent lamps are lighting a propagation table in a glass greenhouse. Even greenhouses need supplemental light.

I use a large eight-lamp HO T5 fluorescent fixture to light and grow a rotating group of houseplants, herbs, orchids, and young transplants.

cost a bit more than the other types and can be used only in ballasts designed to hold HO T5 tubes, but because of their high outputs of light at a low usage cost they are very popular for starting seeds and growing microgreens, other leafy greens, herbs, and houseplants. Most of the HO and advanced-spectrum fluorescents you will want are T5 tubes.

Many home-improvement stores will carry a bevy of basic fluorescent lamps, and some may stock fluorescent grow lights. But don't rush to the hardware store and grab just any fluorescent tube. You should distinguish between fluorescents used for ambient room lighting and those designed for plant growth.

Traditional fluorescent lamps, both for ambient and grow lighting, typically skew to the blue light spectrum. While the cool light makes them effective for growing leafy greens or plants in their vegetative stages, it is not as ideal for forcing flowering plants and heavy-fruiting crops. Advances in fluorescent grow-light technology have resulted in more spectral choices. Nowadays, fluorescent grow lamps are available in distinctively warmer spectrums, as well as full-spectrum forms that provide a better balance of light across the entire PAR light spectrum, which allow you to grow plants more effectively through their vegetative and flowering phases. Certain fluorescents even provide narrow spectrums of light—focused in the upper red, blue, pink, and even green spectrums—plus UV radiation. These narrow-spectrum fluorescents allow you to easily trigger or maintain specific growth and development phases.

CLOCKWISE
FROM TOP LEFT

You can use any sort of fixture for your fluorescent lamps—shop fixtures are just fine—but it must fit T5 or HO T5 lamp tubes.

I mix different color spectrums with different lamps in this large HO T5 fixture, depending on which plants I'm growing.

Some fluorescent grow lamps are available with spectrums weighted to specific color ranges, such as red or blue.

Single-bar fluorescent grow lamps with reflectors can be hung or mounted, horizontally or vertically, to different surfaces.

GROW LAMPS

I designed these custom kitchen grow shelves to keep fresh herbs handy. I had an electrician run the lamp wires into the wall to connect to power and mounted a light switch to turn them off and on.

OPPOSITE TOP Two- and four-lamp HO T5 fluorescents are good options for seed starting.

OPPOSITE BOTTOM I use this large adjustable shelf with four-lamp HO T5 fixtures to start seeds, grow microgreens, and transition transplants. Large fixtures can get pretty warm, so you may not need a seedling heat mat. You can also remove one or two bulbs to reduce light and heat load if your young seedlings seem to need a bit less of either.

In a fluorescent fixture that holds four to eight tubes, you can use a combination of lamps with different color spectrums. If you buy a T5 fixture that holds eight fluorescent tubes and is zoned with multiple on and off switches, you can turn on only your blue lamps during the vegetative phase and then switch on the red with the blue to initiate flowering.

Fluorescent fixtures are relatively small, and you can mount them to all sorts of structures. If you have only a small space on the kitchen counter under the cabinets, you can mount an individual fluorescent fixture for one T5 tube right there. This is a perfect spot for a few leafy kitchen herbs, such as basil or chives.

Many countertop setups allow you to grow small edibles, such as microgreens, small lettuce transplants, small herbs you keep snipped, and seedlings you plan to transplant outdoors. They typically take up only 2 to 3 square feet of space, and they may fit under your kitchen cabinets. If you want to keep fresh herbs handy in the kitchen while you cook, these are great indoor-growing setups. It's hard to get much more hyperlocal. Of course, you can always tuck these small units into any other spare spot in your home. Just remember that as plants grow and reach the lamp they may burn, so harvest herbs and transplant seedlings before they get too tall.

Small fluorescents don't generate as much heat as incandescent lamps, so you can situate them much closer to your plants, especially young seedlings, for better results. A smaller two- to four-lamp fluorescent fixture can sit just a few inches above young seedlings. As plants grow larger, move the fixture a foot or so higher. Large fluorescent fixtures may need to be 2 to 3 feet above plants to avoid causing heat stress.

Compact Fluorescent Lamps
A CFL is a good option if you want the efficiency and effectiveness of a fluorescent lamp but prefer to use a spotlight fixture, track lighting, or other directional lamp for a group of houseplants or a single large houseplant. CFLs look like a twisted or folded-up fluorescent tube. Many CFLs do not require a special ballast, so you can simply use them in your standard light fixtures such as ceiling, spotlights, or floor lamps. The light they produce is acceptable in your open living spaces. CFLs are a good lighting solution for smaller spaces, small grow tents, and areas where you do not want too much heat buildup. Most CFLs are

A close-up of a large CFL.

full spectrum, leaning to the cooler side, but you can also find them with a warmer Kelvin color-temperature rating.

Just as with T5 fluorescents, you will have to place the CFL fairly close to your plant so you don't lose light volume. Because of their moderate heat output, this is relatively safe. This feature also makes them easy to hang horizontally or vertically if you need to get more light at lower levels between large plants. If you have big plants or an expansive grow area, or are growing inside a grow tent or a closet, you can use larger CFLs with light reflectors to spread the light over a larger area and quantity of plants.

Large CFLs, such as those in the 250-watt range, will require a special fixture and reflector that can accommodate the larger base on the bulb.

I don't grow heavy-flowering or fruiting plants, such as tomatoes or squash, to fruiting maturity using CFLs, but I often use a 250-watt CFL for growing plants in a vegetative stage, as well as for young tomato or pepper transplants, leafy greens, leafy herbs, foliage house plants, and maintaining orchids and other tropicals when out of bloom. If you have a collection of houseplants, orchids, or succulents that need additional light, CFLs are a good choice.

HIGH-INTENSITY DISCHARGE LIGHTING

Traditionally HID lamps have been the standard for serious amateurs, professional growers, and large-scale plant-production operations. To the new indoor grower starting small, HIDs may not seem like the right fit. The bigger price tags associated with HID lamps can be discouraging. However, many serious growers will

You can use a CFL in household lamp fixtures and as grow spotlights for small indoor houseplants or cuttings.

LEFT This hanging lamp with a small CFL illuminates a potted gardenia.

still argue that today's HID lamps are a better value than fluorescent or LED lamps because they yield higher levels of light output for the input costs. Cramming your space full of extra fluorescent or LED lamps in an attempt to increase light output (which you could get from fewer HIDs) could cost more and generate more unwanted heat than if you used a good HID lamp.

The enclosed spaces in which I use CFLs can get almost as warm as with high-pressure sodium (HPS) lamps of an equivalent wattage. But I get more PAR from

GROW LAMPS

HPS lamps. If you are growing large groups of plants indoors, growing heavy-flowering and fruiting plants, or growing in a large space, HIDs are the typical choice. If you are growing smaller groups of leafy greens, herbs, or ornamental plants that do not need intense light levels to flower, you may not need HIDs.

There are a few types of HID lamps from which to choose, mainly metal halide (MH) and HPS. HIDs differ in terms of energy usage, longevity, and output, so you need to weigh the pros and cons of each before investing in a quality lamp.

Metal Halide Lamps

MH lamps are created using a few types of gases, the most common of which is sodium iodide. They generally emit a light spectrum range that closely mimics that of natural sunlight, weighted to the blue and violet spectrum, so they are good for all vegetative stages of plant growth. The light produced from MH lamps looks white and bright with a cool visual temperature. While you can also use MH lamps during the flowering stages of growth, many growers employ them only when growing seedlings, cuttings, or vegetative transplants, or bulking up vegetative growth on larger plants. Afterward they switch to a warmer-spectrum lamp or add in more orange or red light with an additional lamp during the flowering phase.

MH lamps produce up to five times the light you would receive from an incandescent lamp, and can last around 15,000 hours. They use a good amount of energy, however, and require a special ballast to regulate energy draw. They may also cost more than other lamps.

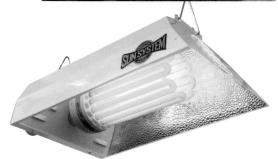

TOP I converted an unused closet into a small grow room for young vegetative transplants, leafy herbs, houseplants, and out-of-bloom orchids. The 250-watt CFL works well in this small space to maintain these types of plants.

BOTTOM A large 250-watt CFL with reflector hood.

Ceramic Metal Halide Lamps

Ceramic Metal Halide (CMH), also known with a branded name of Light Emitting Ceramic (LEC), is the next evolution of MH lamps. CMH lamps use ceramic, just like HPS lighting. Newer CMH lamps require less electricity to generate more usable light output, so you can use a lower-wattage lamp than you might with an HPS or an MH lamp. CMHs are thus more efficient than standard MH lamps, and they put out less heat than HPS lamps, which is a great benefit to most home growers.

CMH lamps offer a solid full-spectrum output, including some UV and far red light. They are available in two Kelvin measurements: 3100K and 4200K. The 3100K lamps are good for growing plants from vegetative to flowering to fruit

Metal halide lamps
produce a cool, blue-
tinted white light.

without switching lamps, whereas 4200K lamps are best for propagation and vegetative growth—you can use one 315-watt LEC to replace two 2-by-4-foot HO T5
fluorescent fixtures. You can also use CMHs as an efficient way to supplement
other HID lighting, such as HPS lamps. Fruiting crops like cucumbers, squash,
peppers, and tomatoes flower very well under a 315-watt 3100K LEC lamp without
any additional red-light sources.

Some CMH lamps come equipped with a built-in reflector and ballast, giving
you an easy plug-and-play option that takes up less space. Some of these complete
CMH rigs are affordable and produce results comparable to more expensive HID
lamps. CMHs are also a good option if you have limited amps or older wiring and
are concerned about your power draw. I have had success using a 315-watt LEC
in a 4-by-4-by-6-foot grow tent for both vegetative and flowering phases with good
regulated temperatures, even without supplemental cooling. Normally, you would
need a lamp that pulls at least 400 watts to adequately light the same space, but
the lower-watt CMH does the job with about the same heat output as a 400-watt
HPS lamp. With a more full-spectrum light output and a 20,000-hour life span, a
CMH lamp is a quality all-around HID lighting solution for home gardeners who
want to grow a variety of plants indoors without switching lamp types.

CLOCKWISE
FROM TOP

CMH lamp bulbs are small but powerful.

Peppers and tomatoes growing and fruiting nicely in a grow tent with a 315-watt LEC lamp.

A plug-and-play fixture that includes a reflector and a ballast for a CMH lamp.

Be sure to read the instructions on the MH or CMH lamp packaging. The lamp may require a 10-hour run time before you can restart it without several minutes of powering down. Turning the lamp off and on again immediately before it has run 10 hours can damage the ignition system.

High-Pressure Sodium Lamps

HPS lamps were the most common type of plant lighting in greenhouse growing operations until LEDs gained popularity. They are used primarily for supplemental lighting where natural sunlight (more blue light) is also present, such as in a glass or poly greenhouse. HPS lamps are also used to extend photoperiod. Overall, HPS lamps are longer lasting and more efficient than MH lamps when you consider their PAR output. Many HPS lamps outperform LED lamps when it comes to plant productivity. Their light spectrum is heavily weighted to yellow, orange, and red, with only very small amounts of blue light, which is why they are used for supplementing ambient natural daylight. If you are looking to boost flower production, HPS lamps traditionally outperform both MH and fluorescent lamps. In terms of total light output, an HPS lamp will provide about six times the light of an incandescent lamp.

If you are considering using an HPS lamp exclusively in an enclosed space that does not receive any natural light, your plants may tend to elongate with thinner stems in a shade-avoidance response. If you are growing in an enclosed space—such as a closet, grow tent, basement, or sealed grow room—exclusively under HPS lighting, plants might get leggy and can become deformed.

If you want to take advantage of the efficiency of HPS lamps but don't want to compromise on light spectrum, you can (1) look for HPS lamps with enhanced performance in the blue spectrum, which are a great compromise, or (2) supplement your HPS lamp by combining it with some blue light from an MH or a CMH lamp, cool-spectrum, narrow-spectrum blue fluorescent tube, or blue

TOP An HPS lamp produces warm-colored light.

BOTTOM This HPS lamp lights up a converted grow closet.

Tomatoes and peppers growing under the yellow-colored light of an HPS lamp.

ABOVE RIGHT The bean plant on the right was grown from seed under only HPS lighting; note its very elongated internodes. The plant on the left was grown under a CMH lamp.

LED. If you are growing in an open space where natural outdoor light is present, such as a greenhouse or a room with windows, you can use HPS lamps alone to supplement the light spectrum and photoperiod—although the orange visual color of the light is less than aesthetically pleasing.

Common HID lighting techniques include growing plants in their vegetative stage under cool-spectrum MH or CMH lamps and then switching to warm-spectrum HPS lamps when it is time for them to flower.

Double-Ended High-Pressure Sodium Lamps

Double-ended (DE) bulbs and fixtures are the next-generation evolution of HPS lamps. DE HPS lamps were originally designed to be compatible with traditional DE MH fixtures. Conventional single-ended (SE) HPS lamps screw into a fixture on one end, like most household bulbs. A DE HPS lamp connects to a fixture on both ends, similar to fluorescent tubes. DE bulbs do not have a metal frame inside the glass bulb, so they have a significantly thinner profile than SE lamps. Their symmetrical form, plus the lack of light-blocking metal frame, enables them to deliver more light, more evenly, to your plants than SE HPS lamps.

DE HPS lamps also have a longer life span than SE lamps. Manufacturers claim that DE lamps produce 10 percent more PAR than SE HPS lamps, and that they also provide more UV and IR light.

The one caveat with DE HPS lamps is that they are filled with nitrogen, which conducts heat from the outside into the lamp. Do not blow cooled air directly on the lamp or use an air-cooled reflector that passes air directly over the bulb itself. Special air-cooled hoods for DE lamps pass cooler air through a chamber above the lamp so as not to impede its functionality. You typically need to place DE HPS lamps a bit higher above your plants than other HID lamps to prevent heat damage.

Dual Arc Lamps

Dual arc lamps are hybrid HID lamps that blend the benefits of the cooler MH light spectrum with that of the warmer HPS lamps. For example, a 1000-watt dual arc lamp combines a 600-watt MH and a 400-watt HPS lamp within the same bulb. Thus, you get a bit less of each type of light emitted in each spectrum in the blended lamp, but the mix of warm- and cool-spectrum light means you don't have to switch between lamp types. You can grow your plants continuously through vegetative and blooming stages under one type of lamp. Manufacturers claim you get more compact plants but still achieve good flowering. So why wouldn't everyone use dual arcs? As of this writing, they are available only in 1000-watt lamps, which can limit their use to gardeners with a larger and more intensive growing capacity. Also, some growers feel that because each part of the lamp emits smaller quantities of light, yields may be lower than with other HID options.

Hyper arcs are the emerging evolution of SE lamps. They generate high outputs of PPF and PBAR and are purportedly some of the most powerful grow lamps on the market. These high-wattage lamps require more than 1000 watts, however, so they are for more serious growers who can provide the necessary space and temperature control.

ABOVE A DE HPS lamp.

BELOW You can see both an MH and an HPS lamp in this dual arc bulb.

Conversion Bulbs

Conversion bulbs operate on the opposite type of ballast they would originally use. For example, a conversion HPS lamp runs on an MH ballast, and a conversion MH runs on an HPS ballast. These are convenient when you want to switch lamp types without buying a new fixture or ballast. Just make sure the wattages match up.

As a safety precaution, remember that fluorescent, CFL, MH, CMH, and HPS lamps contain mercury. Always turn off power and allow the lamp to cool before handling. If you break a lamp, follow the manufacturer's instructions for safe cleanup.

INDUCTION AND PLASMA LIGHTING

Other new kids on the HID block include magnetic induction and plasma lights, or light-emitting plasma (LEP). Magnetic induction lamps work the same way a fluorescent tube does, but they don't have a filament. LEPs have no electrode or filament, but they use sulfur plasma with microwave radiation. Manufacturers of induction and plasma lights claim they are as energy efficient as fluorescents and produce moderate to almost no heat outputs. The realities, however, can vary.

Magnetic induction grow lamps have become popular with cannabis growers because of the spectrum and amount of light they emit, with good efficiency.

LEP lamps are emerging grow-light technology.

The tubes are typically round or rectangular. One side may produce a blue spectrum of light and the other side, red. Because they work like fluorescent lamps you can use them in much the same way, but you can also get a larger quantity of light consistently, and for a longer time period, than with fluorescents or CFLs. Because there is no filament, which is the easiest component to damage and usually the first to die, they last much longer than other types of lamps.

LEP is a relatively new grow-light technology. Plasma lights themselves are not new; Nikola Tesla invented the early versions. LEP lamps operate like a hybrid between HID lighting and a solid-state LED light. They use gases such as halides, sulfur, sodium, and mercury. These gases are excited using either radio frequencies or microwave radiation. LEPs can deliver a wider spectrum of light than traditional MH or HPS lamps, with better efficiency and potentially lower heat output—although the lamps typically need a fan. With manufacturer claims of anywhere from a 30,000- to 50,000-hour life span, over the long term they could be much more cost effective than other lighting choices.

LEP lamps produce electromagnetic interference (EMI, also called radio-frequency interference [RFI]), which means they can interfere with digital ballasts or radio signals. And not all growers agree on the benefits of LEPs or their efficiency. Stay tuned for potential advances and new options in this category.

Many indoor growers are taking advantage of the advances in both fluorescent and LED technology to move away from more expensive traditional HIDs. There is a lot of variability in new lighting options, however, and some experienced growers still insist HIDs provide better results. As horticultural lighting technology advances, HID lighting will likely become even more efficient and effective.

LED LAMPS

The world of LED technology for growing plants has exploded over the past few years. Their growing popularity is based on improving cost and energy efficiency. Most of the self-contained growing units on the market, such as those intended for growing greens or herbs on your kitchen counter or even integrated into your kitchen fixtures, use LEDs.

While LEDs did not start out as the best lamps for growing plants, and there is still a lot of variability among the options available, they are earning a good reputation as low-energy cost and low-heat grow lights. While some indoor growers still use LEDs only as a supplement to other forms of HID lighting, newer advances are making LEDs a good consideration for primary plant lighting, especially in intensive indoor food-production operations that use vertical farming. You can also use LEDs to lengthen photoperiod during the shorter and darker days of winter, or shift easily between vegetative and flowering stages of plant growth using spectrum-specific LEDs.

Small LED lamps and fixtures are typically very affordable. But the larger multiband rigs that emit enough light for multiple plants in a grow tent can cost just as much as, if not more than, HID lighting, even if they do not emit as much PAR.

An LED lamp is a semiconductor that produces light when electrical current passes through it. This is solid-state lighting, as opposed to lamps that use electrical filaments and gases to produce light. This solid state can make LEDs a longer-lasting, more efficient choice. LEDs are lightweight and don't emit a lot of heat, which is very important in a controlled growing environment.

As LED technology advances, so do concepts in indoor plant lighting as decor.

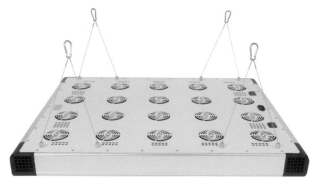

LED lamps produce only about a quarter of the heat that other high-intensity lamps emit because they can convert more watts into light. Less heat means less transpiration in your plants, which cuts down on watering. Low heat output also means you can place LEDs close to your plants without the risk of scorching foliage or damaging young seedlings. When you are growing plants that need a more concentrated delivery of light (such as seedlings), the closer you can place your lights without burning them, the better. But keep in mind that larger LED rigs can still generate heat, especially in enclosed growing spaces.

While the low energy usage and heat output of LEDs are great, the ongoing concern for growing plants relates to both the amount and kind of light (PAR) they can deliver to your plants based on their electrical draw. To date, LEDs are not as efficient for growing plants as other types of lamps. Newer LED lights, depending on the wattage, claim to be just as intense as traditional HPS lights in terms of the amount of PAR they can produce. Real outputs will vary among different manufacturers, and quality construction matters. LEDs may draw less wattage than they claim on the label, so their output of light may be lower than you expect.

Dual-band LEDs that mix primarily red and blue light produce light that looks pink to purple in color. You can use this mixture for continuous growth through vegetative and flowering stages. You can also use single-band nanometer-specific blue LEDs for vegetative growth, then add red if your plants need to flower. Or bulk up young vegetative plants using single-band red LEDs, then add in blue light to encourage flowering.

As LEDs have evolved, they came to include deep red and royal blue light, and were referred to as three- or four-band lamps. LEDs can include multiple bands of light within the PAR spectrum, including green, yellow, orange, violet, and others, as well as UV and IR. These types of multiband LED fixtures are often

TOP Plants growing under an LED fixture in a home grow tent.

BOTTOM Lower heat output from the back of this LED fixture means less heat directed to the canopy of your plants.

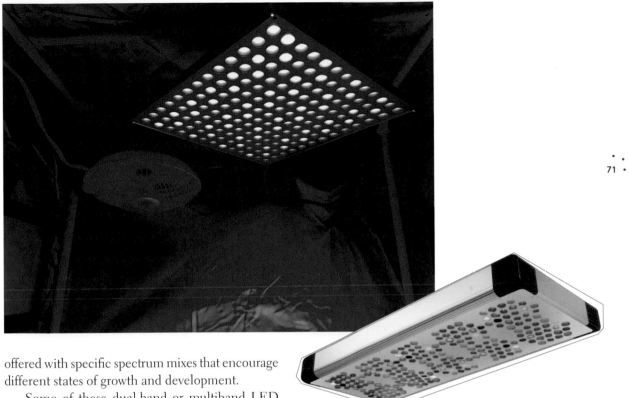

offered with specific spectrum mixes that encourage different states of growth and development.

Some of these dual-band or multiband LED fixtures are zoned so you can turn on only the color or mix of colors you want, depending on your plant's stage of growth. This is handy when you have a small growing space and cannot move plants to different lamps between the vegetative and flowering stages.

White full-spectrum grow LEDs are not really white; they are blue diodes coated with a yellow phosphor, which creates the appearance of white light. Over time this phosphor coating can degrade, and the color of light it produces will change. You can check the spectral output chart on the lamp, if provided. Full-spectrum LEDs will usually lean to the cool or the warm side, and you can use them accordingly for either foliage or flowering plants.

Multicolor white LEDs, also known as RGB LEDs, mix separate red, blue, and green diodes together in a balance to create white-colored light that is both visually pleasing and efficient for plant growth. RGB LEDs are a bit more complicated to manufacture than standard white full-spectrum LEDs, so they may cost more.

ABOVE LEFT This LED fixture mixes red and blue light to produce a pink-colored light.

ABOVE RIGHT This LED rig includes a variety of specific color spectrums.

BELOW A full-spectrum LED bar that produces white-colored light.

ABOVE Use spotlight
LEDs to light individual
potted plants.

RIGHT LED bulbs are
available in various shapes
and sizes.

Many small two-band LED lamps can be used in standard home light fixtures, floor lamps, or spotlights. These small lamps typically fall in the 3- to 13-watt range and will be labeled as E6 or E27 base. LEDs with a GU10 base that snap in are often used in recessed lighting fixtures. These individual low-wattage LEDs are useful for only individual houseplants. Do not expect them to light up a large area or to be suitable for multiple plants.

LEDs lamps are available in other configurations: large multiband fixtures for inside grow tents and grow rooms, light bars you can mount or hang from shelves or cabinets, and even flexible strips you can wind around fixtures you build or the inside of grow tents or closets. You can also use LED strips to build your own LED hanging fixtures. These options give you the ability to interlight plants, or to run lighting between plants to improve growth on the lower areas.

ABOVE LEFT LED clip lights are handy for lighting individual plants on shelves or tables.

ABOVE RIGHT You can hang or mount flexible LED strips to fit areas that are hard to light.

Small self-contained LED units provide both ambient full-spectrum and blue and red spectrum light for plant growth.

T5 LED and Retrofit LED Lamps

Emerging LED technology includes LED bars that are engineered to fit into a traditional T5 fluorescent fixture. Growers replace HO T5 fluorescent tubes with LED bars for better efficiency. These lamps are labeled HO LED bars, which may have to be wired into a fixture, or T5 LED Direct Replacement Bars, which are inserted into the fixture just like a standard T5 fluorescent lamp (they have G5 sockets). If you are currently using a T5 fluorescent fixture, remove the fluorescent tubes and substitute the LED direct replacement bars. T5 LED bars are available in full-spectrum white light, and red and blue diodes for bloom cycles. Newer T5 LEDs that include UV and IR radiation are also available. If T5 fluorescent grow lamps are putting out too much heat for your seedlings or young plants, switch them out for these LED bars.

There are also white LEDs called HID Retrofits, with an E39 base that fits traditional household and commercial high-intensity light fixtures with wattages less than 150. These are generally not manufactured for plant growth, but it is possible to use them.

Given that many LED grow lamps combine only red and blue light, the pink- or purple-colored light they produce might not be the most attractive choice for your kitchen or living room. Pink light also causes your plants to look pink or purple instead of green. This makes it hard to identify a nutrient deficiency, fungal disease, or pest problem. A multicolor white LED, or RGB LED, will give you a more standard white-colored light.

LED grow-light bars (in all white or nanometer-specific colors) are available with small reflectors. You can fit them under cabinets or grow shelves just like the T5 fluorescent units.

Low-heat T5 LED bars that can be used with traditional HO T5 fluorescent fixtures.

You can use LED spotlights to grow individual plants in your living space.

A high-intensity red
LED lamp.

ABOVE RIGHT
A high-intensity blue
LED lamp.

HID LEDs

Some of the newer, larger LED grow-light fixtures, especially those labeled for commercial production, produce a light that mimics the spectrum output of HPS lamps or HO T5 fluorescents, and they are marketed as low-energy replacements for both. These newer lamps may include PAR and PPF measurements, which are handy for calculating their efficiency compared to HID lamps. These large LED rigs run relatively quietly (although they do make noise), might not use as much electricity as HID lamps, may produce less heat (they claim to convert more of the energy into light), and provide a large footprint of lighted growing area. These features make true full-spectrum LED fixtures useful for growing in an open living space where you don't want pink- or purple-colored light. Be prepared for these fixtures to make some noise and carry a heftier price tag than other LEDs.

One limitation of LED lamps is that you cannot replace the bulb once it stops functioning. When a diode quits working, that lamp is kaput. Some manufacturers, however, claim their LEDs will last for upward of 100,000 hours and that you will never need to change a bulb. If the claims are true, you could run one of these fixtures for 16 hours a day, every day, for about 17 years. But in reality, after about 30,000 hours there can be significant degradation in usable light output from LEDs.

The Full-Spectrum Label

Many LED grow-light fixtures labeled as full-spectrum may provide a mixture of only red and blue light, sometimes with IR and UV mixed in. Technically they are not full-spectrum, but nanometer-specific. The visible light these lamps produce has the appearance of a distinct color—blue, red, pink, or purple. This is not a problem for your plants, as they use these spectrums most effectively, but if you don't want distinctly colored light in your living room, these color-specific LED fixtures are better suited for the closet or grow tent.

You can use a full-spectrum LED to design a beautiful and functional indoor growing environment for a beloved plant.

No matter what LED lamp you choose, be at least a bit wary of claims of proprietary spectrums or unique combinations of light colors. If you understand the basics of light science, you know your plants can use all the light within the PAR spectrum, and blue and red light used individually are most efficient for photosynthesis and best employed to shift developmental stages of plant growth.

As LED technology advances, these types of lamps will likely outpace other types of traditional grow lamps in light delivery and energy efficiency. As with any evolving technology, part of the bigger price you pay now (with perhaps less efficacy) is an investment in future advancement and efficiency.

LED grow-lighting options are quickly becoming more creative and versatile.

DETERMINING HOW MANY LAMPS YOU NEED

Much of the information provided online or by manufacturers for calculating your lighting needs is based on watts. While energy use is an important consideration, it is an input, not an output.

If you are not overly concerned about using specific light measurements to generate specific plant yields, you can use basic conversions to ballpark estimates by watts. Without any additional natural sunlight, you can estimate that:

- A 1000-watt lamp lights a 36- to 40-square-foot growing space.
- A 600-watt lamp lights a 20- to 36-square-foot growing space.
- A 400-watt lamp lights a 12- to 16-square-foot growing space.
- A 250-watt lamp lights a 6- to 9-square-foot growing space.
- A 150-watt lamp lights a 3-square-foot growing space.

These light footprints account for only the concentrated growing area, not the surroundings. If you are growing in a sealed space and using reflective wall coverings, you can usually get away with using a lower-wattage lamp. But if you place plants outside these direct growing areas, they may not receive enough light and might stretch.

For example, if you are growing tomatoes in an area of your home or in a grow tent, many sources will recommend about 40 watts per square foot to properly light tomatoes, and a calculation along these lines:

growing area: width × depth = square feet; for example, 4 × 4 = 16 square feet power use of lamp: watts × square feet = desired wattage; for example, 40 × 16 = 640 watts

Using this estimate, you need a lamp that runs on about 650 watts, or two lamps that total the same. But relying only on watts doesn't account for how much PAR a lamp delivers. I successfully grow tomatoes and peppers in a 16-square-foot grow tent using one 315-watt LEC lamp. That's half the recommended wattage, which speaks to the PAR output and efficiency of the lamp, versus just the wattage it pulls.

Peppers growing in
a 4-by-4-foot grow tent
with a CMH lamp.

While you may use watts per square foot to give you a rudimentary idea of how much lighting you need, don't rely on it as a measure of overall effectiveness. Remember, a lamp that uses a lot of power does not necessarily put out a lot of usable light, potentially leaving you with a hefty electric bill and wimpy plants.

If you have the patience and aspire to big plant yields, you should calculate the number of lamps you need based on the PAR they deliver over a given area and time. But you also need to consider that different plants need different quantities of light. Use the target PPF emitted by the lamp and the PPFD target value for the types of plants you're growing, which you can obtain from the plants' DLI requirements.

CALCULATOR ALERT

Number of lamps = (PPFD target × total plant canopy area m²) ÷ (PPF per lamp × CU)

CU is the somewhat ambiguous percentage of light degradation that will most likely occur. Not all the light produced from the lamp will strike your plant canopy; some photons will bounce into other parts of the room. You can use a ballpark CU of 80 percent (.80) or, if you want to be more conservative, 70 percent (.70).

For example, the total number of lamps that each emit 200 µmol/s needed to provide an area of 5 m² with 200 µmol/m²/s at a CU of 0.80 will be (200 × 5) ÷ (200 × 0.80) = 6.25. If you round up, that comes to seven lamps, which may be more than you anticipated. You could bump up to more powerful lamps with higher PPFD values if you want to purchase fewer lamps for the same space. If you are using several light fixtures in a grow room, space them as evenly as possible to equalize light distribution and keep plant growth consistent.

If you want to get into the nitty-gritty of how much it will cost to operate your lighting system, grab your most recent electric bill and locate your kilowatt-per-hour rate. If you pay $0.08 per kilowatt, it will cost $0.08 to run a 1000-watt system for one hour. If you pay more (say, for wind-only energy), you might be closer to $0.15 per kilowatt, or $0.15 to run a 1000-watt lamp for one hour. But you may have some energy loss with your fixture, so your lamp might consume more energy than the label states. You can ballpark an additional 8 to 10 percent additional power usage.

BALLASTS

Ballasts are a key component for all plant lighting. They regulate and balance the electrical current that flows through your lamp from the power outlet and convert it to the proper voltage the lamp requires. Most fluorescent, HID, and LED lamps require a ballast to protect them from electrical currents. Using a ballast will both protect your lamp and extend its operating life; the lamp can be damaged or destroyed without one. Some small lamps can use passive ballasts that don't require additional electricity to operate. High-intensity lighting, however, will typically require a powered ballast. Some fixtures have built-in ballasts, while for others, a special ballast is sold separately from the lamp or bulb.

When you turn on a grow lamp—for example, with a fluorescent tube—the ballast will first deliver a burst of high voltage to create a power arc between the two electrodes on each end of the tube. Once the arc is created, the ballast will reduce the current and voltage to keep the flow and light consistent. The ballast helps your lamp deliver consistent amounts of light over time and regulates its energy efficiency.

Most stand-alone ballasts come in a metal box frame and can be plugged into standard 120V outlets in your home, basement, or garage. Plug the ballast directly into the wall outlet, then plug your lamp fixture into the ballast. It is important to match the ballast to the lamp. If they do not coordinate, you might burn out your lamp. Every lamp has a rating that corresponds to a specific type of ballast. Check this before you buy.

The differences among ballasts relate to the frequency output they deliver to the lamp and how they convert energy into usable light. There are three categories of ballasts: electromagnetic, electronic, and digital.

Electromagnetic Ballasts

Growers have used electromagnetic ballasts, also called magnetic ballasts, for some time with plant lighting, typically for HPS and MH lamps. Most magnetic ballasts allow you to switch between HPS and MH lamps for different stages of growth. You can also use dual arc lamps with standard HPS magnetic ballasts. Electromagnetic ballasts control the current into your lamp, but they don't regulate the frequency, so fluorescent lamps will flicker. Magnetic ballasts are wattage specific, meaning the watts for your lamp and the watt ratting for the ballast must match. These are also very heavy and generate more heat than other ballasts, and they generate a significant hum or buzz, which may disturb sound-sensitive indoor gardeners. On the plus side, magnetic ballasts have simple engineering, so they work with just about any HID lamp (as long as the wattage matches) and you can repair them fairly easily. Magnetic ballasts do not generate RFI, which can disrupt the functioning of other electronic devices as well as cable and internet service.

Electronic ballasts are useful for home growers with lower-wattage grow lamps.

Electronic Ballasts

Unlike magnetic ballasts, electronic ballasts regulate both the electrical current and the frequency, thus eliminating lamp flickering. You can use them with various types of HID lamps. Electronic ballasts will also help your lamps run cooler and more efficiently, so a significant percentage of the energy the lamp draws can be converted to usable light. Electronic ballasts are adjustable, so you can use them with lamps of varying wattages, which is quite handy for switching out lamps

for different stages of plant growth. These ballasts do generate RFI, and the interference can extend to your neighbors' homes. If you place an electronic ballast near a TV or internet cable line, you may disrupt the service. Ballasts with RFI Plus are largely silent. These features make electronic ballasts a more common choice for the home indoor gardener, but some very high-output lamps may still require a magnetic ballast.

Digital Ballasts

Digital ballasts are electronic ballasts containing an additional microprocessor that allows them to better regulate the output voltage and current to the lamp.

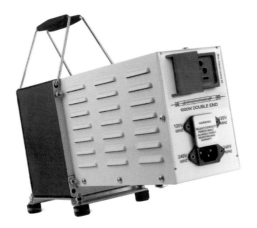

They can also sense when there is a problem or failure with the lamp, and they will not attempt to power a compromised lamp. Digital ballasts can be more efficient than other types while using only a third of the energy. But you cannot use digital ballasts with certain systems, such as plasma lighting. They also emit some RFI, but the technology is improving.

If you are concerned about RFI, make sure you know where the cable lines run in your home and around your neighborhood, and place your digital ballasts as far from them as possible. An easy way to test whether your ballasts are emitting RFI is to grab an old-school AM radio and tune it between stations, where there is no radio signal. Walk around your space and move the antenna around. If you hear a hissing, staticky noise, you have RFI.

Digital ballasts better regulate voltage and current to your lamp.

Plug-and-Play Ballast Fixtures

A great option for most indoor growers is a reflector hood fixture that includes a built-in ballast for your lamp. You can plug in and use these all-in-one options immediately without any setup. These types of fixtures can also be air-cooled: simply insert the lamp into the fixture and plug the entire fixture into a standard outlet. Most T5 fluorescent fixtures are all-in-one reflectors and ballasts. Some of the new HID fixtures also offer this plug-and-play versatility. In terms of both space and aesthetics, it's nice not to have a separate piece of heavy equipment, such as a ballast, you must plug in and make room for. CFLs typically include an integral ballast right in the bulb, so you need only a hood fixture, not a separate ballast.

There are also dual-watt and dual-lamp fixtures that allow you to use lamps of varying wattages and switch between MH and HPS lamps in the same fixture. This is handy if you are transitioning young seedlings, cuttings, or plants in their vegetative phase to mature fruiting plants that consume more energy.

REFLECTORS

Many new growers mistakenly place their grow lights too far from their seedlings or larger plants. The amount of PPFD is reduced the further your plants are from the direct source of light, so your lamp will be less efficient and your yields will be smaller. Fluorescent lamps do not emit a lot of heat, so you can site them pretty close to your plants. When starting new seedlings, you may need to place the lamps within a few inches of the top growth. As plants grow, raise your lamp. If you hang your lights on ratcheted cords or chains, you can adjust them as needed. HID lamps generate more heat, however, so it's not always possible to hang them very close to your plants. Reflectors will help focus light from these lamps where you want it.

Reflector Hoods

Reflectors can significantly increase the efficiency of your overall lighting setup. When you need to disperse light over a larger area, or over a larger group of plants, you'll need a reflector hood. Open-air reflectors help disperse the light from your lamp over a larger surface area. Air-cooled reflectors help control the amount of heat output from HID lighting. Reflectors also help bounce light back down to the lower parts of your plants, which won't get as much direct light as the top foliage. Look for fluorescent fixtures that are lined with a reflective surface to maximize your light from these lamps.

Use lamps that have a built-in ballast with a separate reflector hood to help you best deliver the most light where you need it. You can also use horizontal reflectors that allow you to position your lamp horizontally alongside or in between your plants to ensure an even distribution of light.

ABOVE LEFT An air-cooled fixture for an HPS lamp, which also includes a ballast and a reflector.

ABOVE RIGHT A fixture that allows you to use either MH or HPS lamps of different wattages allows you to switch easily between lamps in the same growing space.

BELOW Use strong ratchet cords to hang your lamps and to raise and lower them easily.

GROW LAMPS

HPS Lamp Distance from Plants

HPS LAMP WATTAGE	DISTANCE FROM PLANTS
150W: not very hot	6–8 inches (15–21 cm)
250W: getting hotter	10–12 inches (25–31 cm)
400W: pretty hot	14–16 inches (35–41 cm)
600W: hot, but most efficient HPS lamp	18–24 inches (46–61 cm)
1000W: very hot	24–30 inches (60–77 cm)

HPS lamps tend to get hotter than other types of lamps, so we will use them as the benchmark for calculating how close you can place your lamps to your plants.

You can place cooler-running lamps, such as fluorescent and LED, much closer to your plants, at about 6 inches, or the equivalent of a 150-watt HPS lamp. But different plants have different heat tolerances, so you may find that your tomato transplants can tolerate hotter temperatures while your lettuce plants scorch. If you notice browning, scorching, curling, or crispy leaves close to the lamp, increase the distance between the lamp and the plant.

Reflective Material

In addition to reflector hoods, you can also employ reflective wall coverings. Reflective material can help you bounce light around all areas of your plants without increasing the number of lamps you use. Many grow tents are lined with such reflective material. There are different types of reflective wall covering, such as Mylar and black-and-white polymer. Note that some reflective surfaces can absorb heat and get hot, so do not let your plants come into direct contact with the material.

**CLOCKWISE
FROM TOP LEFT**

My young naranjilla (*Solanum quitoense*) transplants and brugmansia cuttings are growing about 6 inches below a four-lamp HO T5 fluorescent fixture. As they get taller, I will move them further from the lamps or their foliage will burn.

This tomato has grown about 12 inches too close to the hot 400-watt HPS lamp, and its leaves are beginning to curl because of excess heat. I must raise the lamp to create more distance from the plant.

These peppers and tomatoes are growing several feet below a 315-watt CMH lamp—the perfect distance for this growing environment and these plants.

I covered an unused
closet with reflective
covering to maximize
the lower-intensity
CFL lighting.

Plant-lighting technology is evolving at the speed of light. There may be
more advanced or entirely new alternatives available on the market by the time
you read this book. If you refer to the basics of light science and energy inputs
and outputs, you can apply this knowledge to make good choices, even with the
latest technology.

GROWING CONDITIONS

MANAGING YOUR ENVIRONMENT

Light isn't the only factor that determines your success in growing plants outdoors. You must contend with other environmental conditions, such as temperature, humidity, air circulation, and carbon dioxide levels.

GROW TENTS

If you are looking for a fast, easy way to create a dedicated growing space in your home, basement, or garage, a grow tent is a good solution. Grow tents create a microenvironment in which you have much more control than you would have in, say, an open office or a basement space. You can manage the amount of light delivered directly to your plants, as well as temperature, humidity, and photoperiod. A garage, for example, may offer a lot of room, but it's most likely not temperature controlled and could get too hot or too cold for your plants. Grow tents also make it easier to keep out pests and contain any water or soil spills.

You can situate grow tents anywhere in your home, from your basement to your garage to your home office. They come in many sizes, from tabletop to the dimensions of a small indoor bedroom. If you want a large grow tent, be aware of weight limits if you plan to place it in an upper floor or an upstairs apartment. Once you combine the weight of the gear, containers, plants, and water, the entire setup may be much heavier than you anticipate. And be sure to put down a water barrier in case the tent leaks.

When you purchase a grow tent, you will also need to buy all the other gear that goes along with it, such as grow lamps, fans, venting materials, thermostats, and humidity meters. Some retailers offer bundles or kits that provide all the basic gear. Grow tents are constructed with spaces for you to connect ducts to pull in fresh outside air and ventilate the heat. Some newer grow tents come with built-in fans, thermostats, and other gadgets that help you keep tight control over the growing environment. Ventilation setups range from simple to very complicated. If you are growing vegetables in a tent with a lamp that pulls fewer than 250 to 400 watts, you will probably be able to leave the vents open and use simple fans,

Indoor greenhouses provide the perfect opportunity to grow crops all year long in a manageable environment with total control. Seasonal growing is now obsolete. We create our own seasons in the indoor greenhouse.

modern HOMESTEADING

like two-way fans for a home window, to cool the space. If you are looking to completely seal the tent and prevent any light from escaping, however, and you plan to use a 400-watt or higher lamp, you will need a more formal ventilation setup and artificial cooling that involves intake and output fans with some ducting.

Most grow tent manufacturers provide detailed instructions for setting up the right system for your needs. If you are intimidated by doing a DIY ventilation system, look for one of the complete setups, which can be more user friendly for less-experienced home gardeners. You can also purchase small cooling units that can sit inside a grow room or tent, or external air-conditioners that cool air through a ventilation system, then push it into the tent through ducts.

Modern grow tents are simple to assemble—I have put together several on my own. But if you purchase a tent taller than you are, recruit a partner to help you wrap the skin of the tent around the frame. It's also a good idea to have a second pair of hands help you hang heavy light fixtures, which can be cumbersome. To limit the risk of breakage, wait until the light fixture is secured before you insert any lamps.

Grow tents range in size from small tents that fit inside a home closet to large room-size units.

ABOVE LEFT Cucumber vines growing under LED lamps in a grow tent.

MANAGING YOUR ENVIRONMENT

Figs indoors? Why not? This fig tree is growing under an HPS lamp to encourage flowering.

Note the temperature rating of the grow tent, both for cold temperature protection and heat ventilation. You also need to look for weight ratings for the support structure. You'll be hanging gear from the frame, so it must be able to safely hold the equipment you plan to use. You can buy supplemental supports to hang heavier fixtures.

And, of course, the size of the grow tent will dictate the amount of lighting you'll need to provide. The entire inner surface of grow tents is highly reflective, so you will not lose light to diffusion.

Your setup will depend on your level of interest and commitment, space, and budget. You might use a combination of equipment and setups. I have staged growing areas throughout my home and garage. For edible or flowering plants that need consistently cooler room-temperature growing conditions, I use focused plant lights in living areas, and plant shelves with fluorescent or LED lamps in other spare spaces. Young seedlings, houseplant cuttings, African violets, medium-temperature orchids, and cooler-season vegetable crops occupy indoor areas at any given time year-round. I also have a couple of previously unused indoor closets that work nicely for tucking away crops that like a bright spot in the 65°F to 85°F (18–29°C) range without exposure to very hot or cold temperatures. I like large CFLs or LEDs for these locations. Out-of-bloom orchids, leafy herbs, and vegetable transplants do well in these environments.

In my garage, you'll find several large HO T5 hanging fixtures and full-spectrum LEDs for overwintering large-foliage plants and citrus trees, plus growing herbs, leafy greens, and cut flowers. I also have several shelves with T5s for seed starting and small houseplants. In the summer, I do a lot of vegetative propagation in these areas. In winter, I jump-start seeds of warm-season crops for the winter grow tents and outside garden, and grow extra cold-tolerant plants indoors. (Sometimes I just run out of space in my outdoor garden.)

I also have a couple of tall grow tents in my garage. The flowering tent has HPS lighting, and the vegetative tent uses CMH lighting. Large LED and fluorescent fixtures also work well in these tents.

The grow tents are perfect for winter garage growing. They are insulated, so the HID lamps keep them warm enough during the day and prevent plants from freezing at night. From fall through winter, the grow tents may hold tomatoes, beans, peppers, cucumbers, squash, and other warm-season crops, as well as warm temperature–loving orchids, citrus brought in from outdoors, tropicals I'm coaxing back into bloom, and warm-season herbs.

I live in a hot climate, so my garage gets pretty warm in the summer, even with door insulation. During these months, I either shut down the grow tents or run ventilation fans to cool them.

You can use some lower-wattage lamps in grow tents without cooling them and still keep certain plants happy. It just depends on how big the tent is, how much heat your lamp emits, and the optimal temperature range for your given plants. Simple fans and vents can do the job.

However, when you graduate to higher-wattage, high-intensity lighting, things can get pretty toasty in your closet or grow tent. In these situations, air-cooled lamp hoods are a space-efficient choice. You can hook up these hoods to ducts that pull air from outside the grow room or tent. They directly cool the lamp itself, reducing the amount of heat it puts out in your space.

It's worth the money to invest in a thermometer you can place in your grow

BELOW LEFT Spotlight LEDs keep individual orchid plants healthy.

Seedlings growing on shelves under HO T5 fluorescent lamps.

MANAGING YOUR ENVIRONMENT

ABOVE This corner shelf unit, built out of some recycled-cedar raised garden beds, houses tropicals, orchids and other bloomers that aren't yet flowering, seasonal herbs, and plants that don't have a home.

ABOVE RIGHT Lettuce transplants growing under countertop HO T5 lamps.

RIGHT Two garage garden tents in winter. Tomatoes on the left growing under a warm-spectrum HPS lamp, and peppers, beans, and squash growing under a cool-spectrum CMH lamp on the right.

tent or small grow space to measure air temperature at the plant canopy level. Measuring the temperature regularly will help you make decisions about which types of lamps to use, as well as whether you need to vent and cool your space. It will also help you determine the actual change in temperature from day to night.

TEMPERATURE

When growing plants indoors, it is critical to understand how temperature affects your crops and how to manipulate it to improve your yields.

If you live in the city in a cold climate, you probably crank up the heat through winter months. That heat likely comes from traditional radiators, which are usually positioned under the window—exactly where you put your plants and seedlings in winter so they can get enough light. The result is Radiator-Induced Plant Death (RIPD). All that radiator heat dries out the air around your plants and speeds up respiration.

RIPD

If you have experienced RIPD, you have seen how temperature plays a significant role in how your plants grow, develop, produce, and die. The rate of photosynthesis will increase as temperatures increase, but only to a point. Plant respiration increases rapidly as temperatures rise, as does transpiration. If temperatures become too hot for a plant, respiration can exceed photosynthesis. This means the plant is burning fuel faster than it can make it. If you live in a hot climate, you've likely witnessed this happen to your outdoor plants. No matter how much you water them, they keep wilting or even dying. Their biological functions simply cannot keep up with the temperature. Conversely, temperatures that are too low can slow photosynthesis, shutting down plant growth and flowering. Your goal is to keep photosynthesis running faster than respiration.

Flowering and Flavor

Temperature can trigger flowering in some plants. As temperatures cool, respiration slows and sugar storage increases in some plants. Warming temperatures, after this period of cold, can break the dormancy. Temperature also influences fruit set, fruit ripening, flavor, and even flower color and size.

TOP I can provide optimum temperature and humidity levels for my tomatoes growing in a tent under a CMH lamp.

BOTTOM You can employ high-intensity lighting in small spaces if you use an air-cooled hood.

Depending on the crop you're growing, temperatures that are too hot or too cold can hinder growth and development. Every type of plant has an optimal temperature range for producing healthy foliage and flowers. If you grow a cool-season crop such as spinach or broccoli in temperatures that are too warm or with daylengths that are too long, the plants will bolt. If you have ever harvested bitter-tasting lettuce, temperatures were too hot for your plants. Greens such as collards and kale taste better after they have experienced their first light frost. On the flip side, cool-season crops can come to a standstill in temperatures that are too cold.

If you love orchids, for example, you have probably realized you must choose your orchid varieties by their temperature preferences. It's very tough to grow and flower different species in the same environment when they have individual temperature needs.

Oops. Even professionals forget to water. This dendrobium orchid got too warm and too dry.

Optimal Temperatures

Don't assume, however, that a warm-season crop thrives in hot temperatures. While strawberries typically set and produce ripe fruit during spring and summer months outdoors, they will shut down once temperatures get too hot. Strawberry plants require temperatures between 60°F and 80°F (15–26°C), and they need the night temperatures to be on the cool side. If you grow strawberries in a grow tent that stays at 85°F (29°C), you will have trouble with good fruiting. If you grow your tomatoes or peppers in very hot temperatures, you will have similar troubles: while plants may flower, you might never get fruit set. Outside of optimal temperatures, pollen is rendered unviable and flowers aren't pollinated, which means you don't get tomatoes or peppers. Those green tomatoes on the vine that just won't ripen? Temperatures are either too cool or too hot for the ripening process to work properly.

As a very general rule, warm-season vegetables—such as tomatoes, beans, and peppers—and many ornamental flowering plants grow best between temperatures of 60°F and 80°F (15–26°C). Cool-season vegetables such as kale, lettuce, and spinach, and flowering plants such as cyclamen and snapdragons, grow best at temperatures between 50°F and 70°F (10–21°C). Most plants will grow well at temperatures between 65°F and 75°F (18–24°C), with normal indoor temperature fluctuations. If you are growing a persnickety species of alpine orchid, you may find it tough getting it to bloom in a warm home.

Lettuce flowers can be quite pretty, but they mean your harvesting days are over.

FAR LEFT Broccoli going to flower in warm temperatures. The bees love it.

LEFT Kales thrive in cool temperatures.

ABOVE It's fun to display blooming orchids together, but different species have different long-term temperature and humidity needs.

If your tomato plants keep flowering but don't set fruit, temperatures are too hot.

ABOVE RIGHT If you have tomato fruit on your plants but they just won't ripen, temperatures are either too cool or too hot.

If you have a dedicated grow room or basement, you can install separate temperature controls for your space, but you will still need some venting capabilities. If you're growing plants in a tighter enclosed space, such as a small closet or a sealed grow tent, pay close attention to the heat output of your lamp. Many grow tents specify that, for example, you need to air-cool the tent when using a lamp that pulls more than 600 watts. Otherwise far too much heat will build up inside your tent.

Thermoperiod

A crucial temperature measurement for your plants is the difference between day and night. This change is called the thermoperiod. Tomatoes are a warm-season crop that can grow at temperatures between 60°F and 80°F (15–26°C), but that doesn't mean they will perform their finest at a constant temperature of 80°F (26°C). Most plants do best when they experience a reduction in nighttime temperatures of 10 to 15 degrees Fahrenheit. Plants are adapted to cooler nights in their outdoor environment, so they will perform better with temperature changes that mimic their natural habitat. Plants rest at night in cooler temperatures, when photosynthesis, respiration, and transpiration slow, and recover from stress and water loss. Cooler night temperatures can also increase flower size, keep flower color vivid, and extend flower life.

DIF is the numeric value between day and evening temperatures. You calculate DIF by subtracting the nighttime temperature from the daytime temperature.

When the daytime temperature is warmer than the night, you have a +DIF. When the daytime temperature is cooler than the night temperature, you have a –DIF.

Gardeners can manipulate plant height with a –DIF. Greenhouse growers commonly use a –DIF by lowering morning temperatures by about 10 degrees Fahrenheit in the greenhouse for two or three hours before returning to normal daytime temperatures. This tricks plants into thinking the daytime temperature is lower than at night, which causes plant elongation to slow. For growers who don't want to use chemical growth regulators to keep plants shorter, DIF is a crafty tool.

Dormancy and Vernalization

Growing certain food crops indoors is tricky when you are dealing with a plant that requires temperature shifts, or vernalization. Some plants benefit from a dormancy period, a state of rest induced by a change in temperatures or daylength. Some plants require a vernalization to flower. Be aware that many gardening sites and publications erroneously use the terms *dormancy* and *vernalization* interchangeably, but they are two distinct processes.

Dormancy

The dormancy mechanism helps protect plants during adverse and extreme environmental conditions in winter and summer seasons. While dormant, plants drop their leaves and the growth process slows significantly or stops. After warming temperatures break dormancy, plants continue their flowering process. Florist hydrangeas (*Hydrangea macrophylla*) are an example of a plant that sets microscopic flower buds in summer, or when it finishes its spring–summer flowering cycle; these buds sit undeveloped until after plants experience a dormancy period in winter. Once plants return to warm temperatures and break dormancy, the tiny flower buds will continue to develop. Don't prune this type of plant in spring, or you'll lose your buds. Some bulbs—such as daffodil, amaryllis, and spider lily—will initialize next season's flower buds, go dormant during the heat of summer to conserve resources and avoid damage, and re-emerge the following spring to bloom. The dormant period and cold temperatures can improve flower development.

Some hydrangeas bloom on old wood, holding dormant flower buds through winter; other types bloom continuously on new growth.

Vernalization

For other plants, a period of cold temperatures between approximately 35°F and 50°F (1.5–10°C), calculated as chilling hours, is required before they can develop flower buds or set fruit upon the return of warming spring temperatures and specific photoperiods. This cold-induced initiation of flower buds after exposure to the requisite accumulation of chilling hours is called vernalization. A chilling-hour requirement is a safety mechanism plants employ so they don't produce flower buds too early, potentially losing their flowers (and thus fruit and seeds) to cold or freezing temperatures. The plants keep a memory log of the amount of chilling time they have experienced at the right temperatures to trigger the flowering process. The initiation of the flower buds or bulbs can happen days or weeks after the plants have returned to warmer temperatures and specific photoperiods. Tulips, hyacinth, hardneck garlic, columbine, foxglove, and stone fruits like peaches, apples, and cherries are just a few examples of plants that require vernalization. If these plants don't receive the necessary chilling time, they won't flower, set fruit, or develop bulbs. The winter season outdoors provides these chilling hours naturally (unless you live in a warm climate or experience an abnormally warm winter), while special plant coolers deliver artificial vernalization to certain plants to force them to flower in commercial production.

Hybrid tulips are a classic example of a plant that requires vernalization, as the bulbs will not produce a flower bud until they have been chilled for 8 to 10 weeks at about 45°F (7°C), then exposed to warmer temperatures. You can plant and force tulips indoors, but they will bloom only if they have been vernalized. You can sometimes get away with chilling small quantities of tulip bulbs in your refrigerator before you plant them, but ethylene from the produce in your fridge can interrupt the vernalization and flower bud development.

Hardneck garlic is a bit trickier. It is not only photoperiodic, but it also requires vernalization to form bulbs. This is why you must plant garlic cloves in the fall, then harvest the following spring or summer. You may find garlic a tough crop to crack if you try to grow it indoors at consistent temperatures and photoperiods.

Beneficial Chilling

Some crops don't require dormancy or vernalization, but will bloom better or produce better harvests after they have gone dormant or had a chilling period. Sedum and salvias, for example, don't require cold to flower, but if they get a period of cold they will grow more vigorously and can bloom earlier or better. Short-day strawberry varieties and facultative long-day varieties, for example, benefit from dormancy and some chilling time before you put them in a warm growing environment. Day-neutral strawberries, on the other hand, will flower and fruit continuously without chilling, making them much easier to grow indoors.

ABOVE LEFT If you want to plant tulips in a warm-winter climate, you will need to buy prechilled bulbs and replant new ones each fall.

ABOVE RIGHT I must replant new prechilled tulip bulbs in my Texas garden each December. 'Maureen' is a reliable favorite.

LEFT Garlic needs a specific temperature and photoperiod combination to form harvestable bulbs.

Sedum 'Sea Star' doesn't require any chilling, but it blooms better when exposed to some chilling in winter.

ABOVE RIGHT *Salvia* 'Blue Marvel' sports more vigorous growth and blooms with some winter chilling.

Ultimately, some crops won't be suited to your indoor garden, where you may not be able to properly manipulate their vernalization process or optimal growing temperatures. Best to keep the peach trees and hardneck garlic in the outdoor garden (unless you want to harvest only the garlic leaves). Plant crops with similar needs together. If you are growing in open spaces in your home, your plants will likely have to settle for your human temperature requirements. That might mean you'll have to be patient, as your plants may grow a little more slowly or take longer to produce fruit.

If you live in a climate with a long growing season, mild winters, and very hot summers, you are probably used to growing warm- and cool-season edibles in alternate seasons outdoors, rather than all at once during summer months. Warm-season crops such as tomatoes, beans, and peppers are grown spring through early fall, while cool-season crops such as lettuce, greens, and broccoli are grown fall through winter. To extend your production year-round, you can grow your warm-season crops indoors during the winter and your cool-season crops indoors during the summer.

HUMIDITY

Have you ever struggled with houseplants that turn brown on the tips or edges of their leaves? A lack of humidity indoors can cause many plants to struggle with excess moisture loss. Humidity in the air helps regulate transpiration and water loss or absorption through plants' leaves.

Transpiration

Transpiration is the release of water vapor into the air through plant pores, or stomata, and it serves numerous vital purposes. Transpiration drives the absorption of

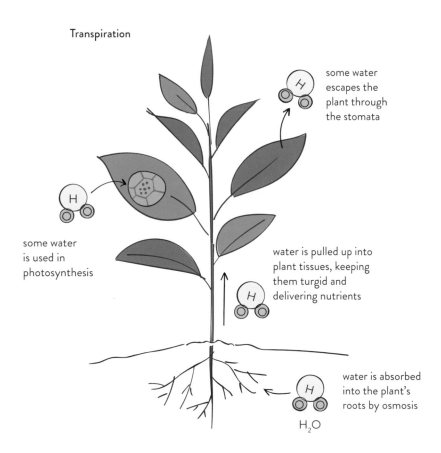

Transpiration

some water escapes the plant through the stomata

some water is used in photosynthesis

water is pulled up into plant tissues, keeping them turgid and delivering nutrients

water is absorbed into the plant's roots by osmosis

H_2O

Transpiration is the absorption of water into the plant through the roots, the movement of water up through the plant, and the release of water out of the stomata.

water up the plant through its roots. It is also responsible for transporting minerals absorbed through the roots into plant tissues, as well as moving sugars generated during photosynthesis. Transpiration keeps plant cells full and firm. This turgidity, or turgor pressure, is what keeps plants upright, pushes new roots into the soil, and keeps plant cells working properly. When plants lose their turgor pressure, they look like they have deflated.

Transpiration is a sort of reverse gravity. Water has a propensity to move from an area of high concentration to low concentration; transpiration is how that water moves in and out of the plant. When humidity is high and there is more moisture in the air, transpiration tends to slow down, meaning less water is pulled up the plant but less water vapor is lost from the plant. When air is very dry, transpiration speeds up and more moisture from the soil is absorbed into the plant and released through the stomata.

Air-Conditioning

Modern air-conditioning units in our homes help cool the air by condensing water vapor and pumping the cooler, drier air back inside. If you run an air-conditioning unit all summer, the air in your home will be drier than the air outdoors. Temperature also greatly affects the rate of transpiration. When air or soil temperatures are too hot or too cold, transpiration slows or even stops. If you live in a very hot climate, your plants are less able to absorb water from the soil and can dry out even if you're watering them during very hot days and nights. Plants in cold climates will stop transpiring altogether when temperatures get too cold. As always, nature seeks equilibrium.

Plants growing in low humidity will dry out faster. When the air around a plant is dry, it creates a high vapor pressure deficit (VPD). Plants that dry out because of low humidity in the air become more susceptible to disease. Pests like spider mites love dry conditions and will attack plants that are already stressed from a lack of moisture.

Tiny seedlings and cuttings are at greater risk for drying out quickly, especially in a dry-air environment. This is why humidity domes are frequently sold with seed-starting kits. It's crucial to keep humidity levels high around your new seedlings or cuttings, as they have a limited root system (or no roots at all) with which to take up water.

Optimal Humidity

Every crop has its own optimal humidity level, which may change depending on the stage of growth. On average, a good target for plant growth is about 50 percent humidity. Some growers will start certain crops at 70 percent during the vegetative stage, when plants are using a great deal of energy and water

ABOVE Without a humidity dome, these tiny microgreen sprouts would quickly dry out and die.

LEFT This basil is extremely wilted because of lack of moisture.

on new green growth, then lower it to 40 percent when plants are putting their energy into flowering or fruiting. But when you're growing in a home with air-conditioning and heating, it's tough to significantly change humidity levels for any length of time. You'll just have to adapt your watering schedule to compensate. If you want to measure the relative humidity in your indoor growing environment, get your hands on a hygrometer. Many thermometers will also measure humidity.

If you're growing in a greenhouse, grow room, or grow tent, you can typically control humidity with a humidifier. Used in conjunction with a humidistat (like a thermostat, but for monitoring humidity), your humidifier can turn on and off automatically to maintain ideal levels. You don't want humidity to be too high, as this can lead to fungal-disease problems on foliage and flowers. You can also use devices that monitor temperature and humidity together.

If you are using an air-cooled lamp hood inside a grow tent and simply allow it to recirculate air from within the tent, it will reduce the humidity in your growing space. It is better to pull in air from outside the tent using ducts to cool your lamp without impacting the humidity. Similarly, if you use an air-conditioning unit in your growing space, you will need to adjust for humidity loss.

AIR CIRCULATION

Beyond cooling and heating the air, you need to keep it moving. Maintaining good air circulation and venting excess heat are especially important in tight, enclosed spaces. Stagnant air breeds diseases, while too much heat damages plant development. When the air in your home, grow room, or grow tent is stagnant, fungal diseases can spread and pest populations can explode.

You may encounter hot spots in your home or grow room. These areas can lead to uneven crop growth and development, and placing an oscillating fan in the space can keep air temperature more consistent and cut down on fungal diseases and pests. Air movement can even stimulate plant growth.

The smaller your growing area and the more tightly it is sealed, the faster air becomes stagnant and heat builds up from your grow lamps. When using a small grow room or grow tent, make sure you have adequate venting capabilities. Heat rises, so outgoing vents should be in the highest place in the room or tent, whereas intake vents should be at floor level. If you're using a tent with an exhaust fan, it may pull enough fresh air into your space to keep temperatures and humidity in balance. Install at least one fan for exhaust and another for circulation. Air should be circulated fully every three to five minutes.

When organic matter is in a confined space, odors can be an issue. If you have a stinky grow room, a carbon filter can help cut down on the smell. Be sure to match the CFM rating on the filter with the CFM rating on your fan.

This small but strong clip fan keeps air moving inside my grow tent.

CALCULATE SIZE OF FAN

When choosing the type or size of fan you need to use in your growing area, calculate the square footage of the space you need to control. Look for the cubic foot per minute (CFM) rating on the fan, and then pull out your calculator:

Length × width × height of your space ÷ 3 (or 5) = CFM to circulate a full cycle of air

This is a basic calculation to help you determine if a fan can circulate all the air in your given space every three to five minutes. However, if you're using intense grow lamps that generate more heat, you can double the CMF calculation to compensate.

CARBON DIOXIDE LEVELS

For most home growers working with typical small edible crops—such as micro-greens, lettuces, and herbs—supplementing with carbon dioxide isn't necessary or practical. If you are using lower-intensity lamps and your growing areas have access to recirculating fresh air, you don't have to worry about CO_2 supplementation. If you're using a grow tent with open vents or external ventilation, supplementation also won't be necessary or very effective. Supplementation will be ineffective if your growing temperatures are below 80°F (26°C).

However, if you choose to grow heavy-flowering or fruiting crops in a tightly enclosed environment—such as a greenhouse, a sealed grow room, or a grow tent—then providing supplemental CO_2 may be beneficial or even necessary, depending on how much usable light the plants receive. The more PAR a plant gets, the more CO_2 it needs for photosynthesis. Plants growing under intense light in a tightly controlled space can run out of CO_2 unless you pump in extra.

CO_2 is a necessary ingredient for photosynthesis, and high or low concentrations in the air around your plants will impact photosynthetic pace. The more CO_2 available to plants, the faster the rate of photosynthesis. Elevated photosynthesis will speed up plant growth and increase yields. Good CO_2 levels also help plants fend off certain diseases.

Plants absorb CO_2 through the stomata in their leaves. For most plants, stomata open during the day, when light-spectrum and temperature conditions are conducive. The longer your plants' stomata are open, the longer they can take in CO_2 and continue photosynthesizing. When the stomata are closed or are not functioning properly, plants do not absorb CO_2. Blue light, temperature, and humidity also influence the opening and closing of stomata.

If your goal is to increase light levels to maximize yields, you may need to add CO_2. Plants growing with enriched CO_2 can also tolerate hotter growing temperatures, which are created when using more high-powered, and intense lighting. Additional CO_2 can then reduce your need to cool the space. If you use external ventilation in your grow tent and still want to supplement CO_2, you can run your exhaust fans at night and your CO_2 generator during the day, when the lights are on.

If you plan on supplementing with CO_2, the amount you will need depends on the type of crop you're growing. Flowering and heavy-fruiting plants, such as tomatoes or cannabis, require more CO_2 than foliage herbs or leafy greens.

Too much CO_2 can cause plant tissues to deteriorate and yields to suffer. In the natural outdoor atmosphere, CO_2 is typically found levels at levels around 300 ppm, which is adequate for most plants. Once CO_2 levels drop below 200 ppm, photosynthesis is constrained. When you boost CO_2 levels to 700 to 900 ppm, crop yields significantly increase. In general, fruits and vegetables need 1000 to

1200 ppm. Plants that produce large flowers or an abundance of flowers all at once, such as roses, require about 1200 ppm. If you're supplementing with CO_2, do not exceed levels of 1000 to 1500 ppm.

At intense light levels, plants will use more CO_2 than at lower light levels. Each plant has its own light-saturation point, at which you can no longer boost the rate of photosynthesis simply by supplying more light. Providing extra CO_2 can raise the plant's light-saturation point.

Air temperature also influences your growing space. As the temperature cools, the CO_2 concentration decreases. This effect is more extreme in tight, unventilated spaces. You can better balance concentrations of CO_2 by raising the daytime temperature in your grow room or tent by 5 to 10 degrees Fahrenheit.

Equipment and ongoing supplies for supplementing CO_2 in your grow room or grow tent can get pricey, depending on the size of the space and how technical you want to get. CO_2 generators, tanks, and regulators are typically involved. This gear can be expensive, and you may need a permit to have it in your home. DIY methods for generating CO_2 include fermentation (creating yeast bins) or using dry ice, but these do not allow you to regulate CO_2 output. The easiest DIY method? Breathing. You can simply take a breathing break in your grow tent a few times a day to naturally boost CO_2 if you don't want to mess with the gear. Not that you have time to stand in your grow tent all day.

COMMON PESTS AND DISEASES

One of the biggest challenges to indoor growers, whether keeping house-plants or growing produce hydroponically, is dealing with insects and diseases. Indoor environments that offer less air circulation, lower light conditions, and less-than-optimal temperatures can be the perfect breeding ground for certain pests.

It's also easy to unwittingly introduce a pest when you bring outdoor plants inside for the winter or buy new plants for your indoor garden. These hitchhikers can quickly reproduce and get out of control.

BIO-SECURITY

Any greenhouse grower will tell you that bio-security is a big deal. Preventing the spread of insects and diseases before they damage plants is always easier than eradicating an infestation or outbreak after it has occurred. In commercial operations, bio-security involves sterilizing boots and shoes before walking into a greenhouse, keeping surfaces clean, sterilizing pots and trays, and protecting water sources.

While you may not go as far as to restrict visitors or make them remove their shoes before they visit your grow room, you can take precautions to limit problems for your indoor plants.

Here are some basic bio-security tips:

- Scout regularly for signs of insects and disease. Treat or remove infested or infected plants immediately.
- Always inspect new plants for any insect hitchhikers or signs of fungal disease. Treat them before you introduce them to your indoor growing space.
- Mind your watering habits. Overwatering is the biggest killer of indoor plants. It can trigger several soil-borne diseases, or simply drown your plants by suffocating the roots. Underwatering triggers stress and certain nutrient deficiencies in plants, which then attract insects like spider mites and whiteflies.

TOP LEFT Aphids on an indoor citrus plant.

TOP RIGHT Scale insects and sooty mold on a greenhouse citrus plant.

BOTTOM LEFT Powdery mildew on zinnia leaves.

BOTTOM RIGHT Not technically a disease, but rather a physical disorder related to water, humidity management, and types of light, oedema blisters can cause defoliation.

TOP Yellow sticky cards are an easy method for catching and killing many flying pests, such as fungus gnats and whiteflies.

BOTTOM High humidity and water on foliage spread diseases such as early blight in tomatoes.

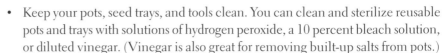

- Keep your pots, seed trays, and tools clean. You can clean and sterilize reusable pots and trays with solutions of hydrogen peroxide, a 10 percent bleach solution, or diluted vinegar. (Vinegar is also great for removing built-up salts from pots.)
- Clean fabric pots by first allowing them to dry completely, then rubbing off any dry debris. Put them in the washing machine or in a soaking tub with a cleanser that contains hydrogen peroxide. Always air-dry fabric pots.
- Between harvests, wipe down the inside of your grow tent or spaces covered with reflective wall covering.
- If you're using a hydroponic system, follow instructions for cleaning and sterilizing, and then be sure to use additives that help cut down on bacterial growth.
- If you have handled infested or infected plants, wash your hands well before you move on to healthy plants.
- If you have pruned an infected plant, sterilize your pruners in a 10 percent bleach solution before you use them on healthy plants. Do not take cuttings from a plant with an obvious infection.

SYNTHETIC VS. ORGANIC TREATMENTS

If you are growing edible crops, be thoughtful about the treatments and chemicals you use. Avoid any systemic insecticide or fungicide on edible plants. Systemic products are absorbed into the internal plant tissues and may persist for long periods to provide ongoing insect- or disease-killing effects. A number of low-impact topical insect and disease treatments are eco-friendly and organic. Always read and research the ingredients on product labels, and never exceed application rates.

SPACING AND VENTILATION

Good ventilation and air movement are necessary to reduce disease pressure. Be sure to use fans in enclosed growing spaces and allow enough space between plants to keep the air moving around them. If your plants are already infected with a fungal disease, however, you may consider turning off the fans temporarily so you don't inadvertently spread spores more quickly.

IDENTIFYING COMMON PESTS AND DISEASES

No matter how experienced you are, it's almost impossible to avoid occasional trouble with certain insects or diseases on your indoor plants. If you grow your plants in soil, for example, you will probably always have some fungus gnats. Don't get discouraged; just watch out for early signs of problems so you can keep them at bay. And if you (or the powdery mildew) kill a few plants, no biggie—you can propagate lots of new ones yourself (see Chapter 7), or you have a good excuse to hit the local plant shop.

Common Plant Pests

PEST	DESCRIPTION	TARGET PLANTS	SIGNS/ SYMPTOMS	TREATMENT
Aphids	Small; oval-shaped; green, black, or white. Aphids suck water and nutrients from leaves and stems. They also carry viruses they pass from plant to plant as they feed.	Annual flowers, beans, beets, bok choy, cannabis, chard, cucumbers, fruit trees, herbs, lettuce, perennials, roses, shrubs, trees.	Curled or wilted foliage, stunted growth. General decline. Affected plants become more susceptible to other pests and diseases.	Insecticidal soap, horticultural oils, spinosad. Wash off with high water pressure, or remove manually (i.e., squish).
Caterpillars	Vary in size, color, and season. Some mature into butterflies, others into moths.	Annual flowers, cannabis, carrots, cole crops, grasses, herbs, perennials, shrubs, tomatoes, trees.	Chewed or curled foliage. Just one or two caterpillars can strip some plants clean of foliage.	*Bacillus thuringiensis* (Bt), spinosad. Trichogramma wasps and other predatory insects.
Fungus Gnats	Tiny black gnats that fly around the potted plant. Small white larvae in the soil.	Any potted plants you're growing in soil that contains organic matter.	Larvae feed on plant roots in the soil. Gnats flying about the plants and room. General plant decline.	Boost air circulation with fans. Reduce overwatering. Sticky traps. Use 10% hydrogen peroxide soil drench, Bt granules.

PEST	DESCRIPTION	TARGET PLANTS	SIGNS/ SYMPTOMS	TREATMENT
Leaf Miners	Beetle, fly, and moth larvae that live in the leaves of plants, feeding on soft tissue.	Beans, berries, cabbage, greens, peppers, and a variety of annual flowers and perennials. Trees and shrubs.	Thin white tracks within the leaf. Healthy plants may not show any stress. Less vigorous plants may have stunted growth and reduced yields.	Improve plant health by amending the soil and proper feeding and watering. Predatory insects, systemic insecticides.
Mealy Bugs	Soft-bodied, wingless, cottony masses, some with long tails. On leaves, stems, bark.	Citrus, foliage plants, shrubs, succulents, trees.	Yellowing and curling of leaves. Sticky honeydew residue.	Wash off, remove manually; insecticidal soap, horticultural oils, spinosad. Systemic insecticide. Predatory insects.
Scale	Oval-shaped insects with either a hard or cottony shell, with different colors. Cluster on stems and base of leaves.	Avocado, basil, citrus, foliage plants, shrubs, succulents, trees.	Stunted growth. Weak appearance, shriveled and yellow leaves that often drop off. Sticky honeydew residue; fungus can grow on honeydew.	Wash off, remove manually. Insecticidal soap, horticultural oils, spinosad. Predatory insects. Systemic insecticide. Can be difficult to treat because of hard outer shell.
Spider Mites	Fast-moving, tiny arachnids with red-brown or pale coloring. Underside of leaves and stems. Fine webbing on leaves.	Annual flowers, basil, berries, cannabis, citrus, foliage plants, herbs, perennials, roses, shrubs, tomatoes, tropicals.	Pale or yellow mottled cast to leaves, leaf curling and dropping. Tiny pinprick holes on leaves.	Insecticidal soap, horticultural oils, spinosad. Predatory insects. Miticide. Persistent and require multiple treatments.

Common Plant Pests *(continued)*

PEST	DESCRIPTION	TARGET PLANTS	SIGNS/ SYMPTOMS	TREATMENT
Thrips	Very tiny winged insects. Wingless larvae cluster in large groups in the crevices of flower petals and new leaf growth to feed on plant tissue.	Annual flowers, beans, carrots, citrus, onions, perennials, roses, squash.	Plant leaves develop a pale cast, then drop. New flower and leaf buds are twisted and deformed as well as discolored. Spreads several viruses and fungal diseases.	Sticky traps, horticultural oils, spinosad. Can be difficult to treat.
Whiteflies	Tiny white flies on leaves and stems. If you shake the plant, they will fly around the plant. Larvae and nymphs suck plant sap.	Cannabis, citrus, cucumbers, foliage plants, grapes, potatoes, shrubs, squash, strawberries, tomatoes, tropicals.	Mottled leaf appearance, yellowing and dropping leaves, overall reduced growth and vigor. Sticky honeydew followed by sooty mold.	Sticky traps, insecticidal soap, horticultural oils, spinosad. Persistent and require repeat treatments.

Common Plant Diseases

DISEASE	DESCRIPTION	TARGET PLANTS	SIGNS SYMPTOMS	TREATMENT
Basal Stem Rot (*Armillaria, Ganoderma*)	A wood-rotting fungal disease affecting a variety of plants.	Coniferous plants, perennials, palms, woody trees.	Fungal conks growing off the base or side of the trunk. Slowed growth and defoliation.	Well-draining soil, planting at proper soil depth.
Early Blight (*Alterneria*)	A leaf spot fungus spread by splashing rain, irrigation, insects, and garden tools. Common in humid, warm conditions.	Eggplant, peppers, potatoes, tomatoes.	Lower leaves develop small brown spots. Spots spread turning the leaf yellow, which then curls and drops.	No overhead irrigation or misting. Water at the soil level. Remove infected leaves immediately and improve air circulation. Foliar fungicide.
Damping Off and Root Rot (*Pythium, Rhizoctonia*)	Damping Off: Fungal pathogens that rot seedlings right at or just below the soil line. Root Rot: Many pathogens that attack root systems, turning them brown.	All seedlings, young transplants. Any plants growing waterlogged soils can suffer from root rot. Can be an issue in hydroponic systems.	Seedlings fall over at the soil line before they can mature. Plants begin to droop, wilt, turn brown, and collapse. Root tissue is brown and can be slimy.	Don't overwater seedlings or keep humidity too high. Use a sterile soil mix and manage soil temperature. Improve soil drainage and aeration. Oxygenate hydrosystems. Add beneficial bacteria.

DISEASE	DESCRIPTION	TARGET PLANTS	SIGNS SYMPTOMS	TREATMENT
Gray Mold (*Botrytis*)	A mold disease that grows on leaf surfaces, blocking light from the leaf surface and causing severe damage to foliage and flowers.	African violets, berries, cannabis, carrots, flowering annuals, foliage plant bulbs, grapes, peas, roses, tomatoes, and many more.	White spots on leaves or stems that turn gray then brown. The fungus can cover large areas, or the entire plant with webbing.	Good soil drainage. Improve air circulation between plants. Remove infected leaves and blooms. Foliar fungicide.
Late Blight (*Phytopthera*)	A leaf spot fungus spread by splashing rain, irrigation, insects, and garden tools. Common in humid, warm conditions.	Eggplant, peppers, potatoes, tomatoes.	Older leaves develop dark green and gray water-soaked spots that grow large and turn brown. Dark blotches on plants stems. Damages fruit and tubers.	No overhead irrigation or misting. Water at the soil level. Plant resistant cultivars. Apply a foliar fungicide. Destroy heavily infected plants.
Powdery Mildew (*Podosphaera*)	A fungal disease with fuzzy white growth. Spreads quickly, covering foliage and blocking photosynthesis.	Annual flowers, beans, cannabis, citrus, cucumbers, foliage plants, herbs, roses, peas, peppers, squash, tomatoes, and many more.	Powdery white substance on foliage. Reduced overall growth and vigor, stunted yellowing leaves that drop.	No overhead irrigation or misting. Control humidity levels. Water at the soil level. Foliar fungicide, sulfur vapor.

DISEASE	DESCRIPTION	TARGET PLANTS	SIGNS SYMPTOMS	TREATMENT
Sooty Mold (*Alternaria, Cladosporium*)	Several types of fungal species—a gray- to black-colored mold that grows on the honeydew residue produced by aphids, scale, and whiteflies.	Any plant susceptible to aphids, scale, whiteflies. Annual flowers, citrus, herbs, perennials, shrubs, trees.	Mold spreads across the leaf surface, blocking light and photosynthesis. Leaves yellow and drop. Overall reduced growth and vigor.	Wipe off plant leaves with soapy water, a plant wash product, or neem oil. Control aphids, scale, and whiteflies.
White Mold (*Sclerotinia*)	A white-colored fungal disease. White fuzzy masses destroy leaves, flowers, and fruit.	Beans, cabbage, chives, lettuce, peas, tomatoes.	Water-soaked spots on flowers, stems, leaves, and fruits. White fuzzy masses. Leaves yellow, wilt and die; fruit is damaged.	Remove infected plants immediately. Plant resistant varieties. Use a foliar fungicide or plant wash as a preventive.
Bacterial Stem Rot (*Erwinia*)	A bacterial disease that enters plant tissue through wounds in their stems and leaves.	African violets, annual flowers, carrots, eggplant, foliage plants, peppers, philodendron, potatoes, tomatoes, and more.	One or two branches wilt first, with water-soaked black lesions on stems. The rest of the plant may then wilt and die.	No effective chemical control. Avoid plant injury. Use clean cuttings and tools; keep workspace clean. Remove infected plants immediately.
Citrus Canker (*Xanthomonas*)	A bacterial disease that causes spots on stems, leaves and fruit of citrus plants.	All citrus family plants.	Yellow to brown spots with a yellow halo on leaves and fruits. Premature leaf and fruit drop. Fruit can still be eaten but will be deformed or unsightly.	Keep your grow space clean, quarantine or dispose of infected plants immediately. Avoid plant injury.

TECHNIQUES FOR ELIMINATING PESTS AND DISEASE

When it comes to pests and diseases, prevention is always best. If your plants do experience an outbreak, here are a few different methods you can use to treat the problem.

Sacrifice the Weak

Cull the herd if you have a pest or disease outbreak among your indoor plants. Consider removing any heavily infested plants in order to limit the impact on less affected plants. The more pests or disease spores present, the more they will reproduce and the harder it will be to eliminate the infestation.

Prune and Clean

When fungal disease or insect egg clusters are present on only a few plant leaves, remove these leaves immediately and throw them away. This will not only reduce the disease or insect pressure, but also increase airflow through your plants.

Biological Control

Predatory mites, lacewings, and ladybugs are a few examples of beneficial insects that are very effective in controlling mites and other harmful insects. If you are pesticide free, you can employ beneficial insects in both outdoor and indoor operations. Do not expect beneficial insects to persist long term in an indoor enclosed space; eventually their food and water will run out and they will die off.

Suffocate and Kill

Insecticidal soaps, neem oil and other horticultural oils are effective for eliminating a number of pests and reducing infection or spread of fungal diseases. They also serve double duty as a natural leaf-cleaning and shine product. Horticultural oils can burn plants when temperatures are hotter than 90°F (32°C), however, so take care in closed grow tents with HID lighting.

Products containing spinosad (a natural chemical found in bacteria) are also highly effective for killing pests. While spinosad can be dangerous for honeybees and other pollinators when wet (spray only at dusk in the outdoor garden), these pollinators will not typically be flying around your indoor garden. Spinosad is nonselective, meaning it will kill most insects on contact. It's a quick knock-down treatment for persistent pests such as aphids, spider mites, and whiteflies.

TOP Sooty mold can completely cover leaves of infected plants.

BOTTOM Mealy bugs cluster together at the ends of stem tissue and will hop away if disturbed.

COMMON PESTS AND DISEASES

Whiteflies are a persistent indoor plant pest on certain plants, but here come the ladybug larvae.

RIGHT Ladybug larvae are ferocious predators of pests like aphids and whiteflies. Ladybugs manage to find their way indoors when my citrus plants have whiteflies.

Thuricide (*Bacillius thuringiensis*, or Bt) is an excellent natural treatment for chewing caterpillars when sprayed in liquid form, and you can use the dried bits in pots to control fungus gnats in containers and mosquito larvae outdoors. While other products are available, these are the most reliable natural treatments for your indoor garden.

Fungal Sprays

A foliar spray with an organic fungicide, a copper fungicide, a horticultural oil, or a plant wash is a good way to combat fungal diseases such as powdery mildew. Potassium bicarbonate is commonly used to control fungal diseases. It works by starving fungal spores of water, which in turn kills the fungus. If you are foliar feeding your plants, stop temporarily in order to reduce moisture on the infected plant's leaves. Foliar feed your plants in the morning so moisture doesn't sit on the foliage all night.

Sulfur Vapor

Sulfur vapor can be an effective way to prevent powdery mildew and other fungal diseases on indoor-grown plants kept in sealed spaces like grow rooms and grow tents. (Do not use sulfur in open areas of your home.) Special evaporators are used with sulfur prills (small granules) to emit a sulfur vapor that controls the pH on the plant's leaf surface. The sulfur vapor particles land on the leaf, then slowly change into sulfuric acid. The resulting change in pH inhibits the fungal spores from germinating. You can run sulfur vaporizers for 8 to 12 hours the first night of treatment, then once a week for 2 to 3 hours per week when grow lights are off, until the disease is under control.

Aphids are common on new indoor-grown pepper plant foliage.

IDENTIFYING THE CAUSE

If your grow room, grow tent, or outside garden becomes infested with insects or disease, there is probably an underlying issue such as incorrect lighting, nutrient deficiencies, or poor maintenance practices. Or you may unknowingly introduce new plants with issues. In addition to wreaking direct damage, insects can also spread fungal and bacterial diseases to your plants. Make sure your plants are getting enough light, you're watering and fertilizing properly, and growing temperatures are in the correct range, and inspect all new plants for insects before you add them to your garden.

Ultimately, growing food and flowers indoors is not a set-it-and-forget-it activity. It requires commitment, time, and practice. You will need to engage daily with your plants to monitor your environment, their needs, and the overall progress. But caring for your indoor plants on a daily basis can become one of your most enjoyable and stress-relieving responsibilities.

PROPAGATION AND PLANT CARE

Now for the fun part: making more of your own plants (because you will always need more plants). There is nothing more exciting than watching your new seeds sprout into baby seedlings or witnessing your cuttings take root. Propagation requires patience, but for true plant lovers, it is a rewarding form of meditation.

Many new gardeners are afraid they will kill their seedlings or cuttings. My answer? Don't worry: you will. First-time failures, especially with seedlings, can make you reluctant to try again. Fear of killing plants may even stop you before you get started. Some people won't buy plants in the first place because eventually they will die, a condition I refer to as Green Guilt. Killing a few plants, however, is ultimately how you learn to grow them successfully. Have faith that there are lessons in early failures, and resolve to learn. Then try again. Green thumbs are earned, not born.

RAISING PLANTS FROM SEED

Sowing new plants from seed is downright exciting. It's also economical. Starting your own seeds is a very cost-effective way to grow a large quantity of plants quickly, plus you'll be able to grow a much wider variety than if you buy only finished transplants from a garden center. If you grow your plants outdoors, starting all your seedlings or cuttings inside can give you a jump on the growing season.

The Right Kind of Light

If you skip supplemental indoor lighting for young seedlings and plant starts, you will usually be disappointed. Most windowsills simply are not bright enough. Provide your seedlings with good lighting to set them up for success from the get-go.

Depending on your climate and outdoor growing conditions through the seasons, you may have to start seeds of certain crops indoors a few weeks, or even months, ahead of outdoor planting days in order to harvest a successful crop. This is where an indoor seeding station comes in handy. You need only a few square feet of space to prepare a couple of seed trays with a small lighting setup.

ABOVE Learning to propagate your own plants is rewarding and fun.

LEFT Young basil, tomato, and lettuce transplants grown from seed under fluorescent grow lamps.

Many seeds require darkness underneath the soil to germinate properly. When germination occurs in darkness, all the seed's energy funnels into root growth. Once the shoot breaks the soil surface and is exposed to light, things change dramatically. Root growth slows and shoot elongation acceleration kicks into gear, with the biological goal of garnering new energy as quickly as possible from photosynthesis. This light-dependent developmental shift is one part of plant photomorphogenesis.

Plants that evolved to have their seeds exposed to the soil surface after dropping are adapted to germinating in the presence of light. Lettuce seeds, for example, need to be exposed to red light or they won't sprout. Sprinkle lettuce seeds on the surface of the soil and lightly press them down, but don't cover them with soil. Do the same with the seeds of many annuals, wildflowers, and perennials. Check the seed packet for information about germination light (or dark) requirements before starting your trays.

TOP Lettuce seeds need exposure to light to germinate.

BOTTOM Once germinated, tiny seedlings need to be close to a bright light source so they don't stretch.

Duration of Light

Once the seedling shoot emerges from beneath the soil, it is not only the presence and intensity of light that matters for good development, but also how long the light is shining. After sprouting, seedlings will need 14 to 16 hours of bright light to grow strong and sturdy. Most indoor rooms, even those with large windows or plenty of diffused light, are not bright enough for baby seedlings. If your seedlings are elongating and stretching, they aren't getting enough light, especially blue light. Stretched seedlings will frequently topple and fail. If an overstretched seedling does survive long enough for you to transplant it into a larger container, the plant often has a tough time thriving and may not produce the desired results or harvest.

Distance from Light

The light source must be very close to your seedlings as they germinate, as near as 3 or 4 inches from the seeds as they emerge from the soil. If the light is further away, young sprouts can quickly stretch beyond

These tiny seedlings are too far from the grow lights above, and they will begin to stretch if not moved closer.

LEFT These lettuce seedlings are growing well just a few inches below the grow lamps.

their ability to remain intact. Ratcheted cords or chains that allow you to move light fixtures up and down will enable you to place your lamps very close to seedlings as they are just emerging, then lift the light source as they grow. If you are using a fixture with adjustable shelves, you can create low and high sections. Start the seeds on a lower shelf close to the light, then move them to higher shelves as they grow.

Grow Lamps

Some lamps generate more light or more heat than others. Fluorescent lamps, CFLs, and LEDs are typically the easiest options for starting young seedlings, as they generate lower levels of heat than HID lamps and you can place them closer to your tiny plants. If you site an HPS lamp a few inches away from young seedlings, the intense light and heat will quickly fry them. Once your seedlings grow up, you can employ more intense types of lighting.

Preparing Your Seeds

Some seeds are more difficult to germinate than others because they have a very hard, protective seed coat. While most annual vegetable seeds do not require any special preparation for germination, seeds of some natives, perennials, and fruits require a bit of extra work, whether soaking, scarification, or stratification.

For most annual and edible seeds, normal germination occurs at optimal soil temperature and moisture levels without any special techniques. You can speed up the germination process (or improve germination rates from older seed stock) if you presprout them, a process called chitting or greensprouting. Chitting involves soaking the seeds, usually for 24 hours, before you sow them. Moisten some dish towels, paper towels, or newspaper, then set the damp material in a tray, on a plate, or inside a plastic bag. Spread your seed onto the moist surface. Seeds will absorb the moisture and swell, and some will germinate and sprout. Sow these sprouted seeds immediately. Chitting your seeds can give you a jump-start on germination time and may allow you to start seeds in cooler-than-optimal soil. Chitting is easiest with larger seeds that you can handle easily (beans, corn, peas), but you can do it with smaller seeds too.

Seeds with very hard seed coats may require scarification to germinate. A seed has a hard seed coat if no amount of squeezing or tapping on a hard surface produces any give. You may have discovered your seeds have a hard seed coat because they never germinated or took an extensive amount of time to germinate. Information on the seed packet usually includes the type of seed coat and any special germination needs.

Seed scarification involves scraping away part of the hard coating to expose the seed to water and gases that trigger germination. In nature, temperature, soil microbes, and even fire can break down seed coats. Animals eat seeds, which are

Borage plants are prolific seed producers.

LEFT Presoaking certain types of seed can speed up germination and improve germination rates.

then exposed to stomach acid that breaks down the seed coat. The bigger the seed, the more likely it will need scarification to germinate successfully.

If you have had trouble getting certain types of seeds to germinate and they have a very hard outer coat, try scarifying them. There are a few easy ways to do so.

Hot water Boil water and pour it into a bowl. Let it cool for about 30 seconds, then add the seeds. Allow the water to cool to room temperature and the seeds to swell, then remove them. The amount of time needed for seeds to swell will depend on the plant variety. It may take a few hours or a couple of days. Soaking for 12 to 24 hours is recommended, but don't let seeds soak more than 48 hours.

File Use a small file or sandpaper to rub away or nick a small section of the seed coat. (Do not rub off the entire coat.) Make your mark on the opposite side of the seed-eye, where the shoot will emerge.

Sand Roll or rub your seeds in or with sand for a few seconds. You can do this between sheets of paper or another flat surface. Don't crush the seeds; just scrape away a bit of the surface coating.

Some seeds do not have a hard seed coat, but rather go into a dormancy period that can make germination a long affair. You will have to wake these seeds from their slumber using a process called stratification. Many natives, wildflowers, herbaceous perennials, and grasses fall into this category.

Stratification involves creating favorable environmental conditions in which the seeds would germinate in nature. Seeds and plants that go into dormancy typically come to life after the return of spring temperatures and rainfall. You must mimic the cycle of dormant winter season followed by natural germination season. For starting seeds and growing indoors, create an artificial winter by chilling the seeds.

Some seeds experience a double dormancy, so they will not germinate or put on green growth above ground until they have experienced two winter seasons. Species in the lily family require a double dormancy. To speed up production, trick these species by using a repetitive stratification process.

There are two common methods for seed stratification: moist and dry. For moist stratification, mix your seeds with damp sand and seal them in a container. Store the container in your refrigerator for one to two months. Remove the seeds

Wildflowers such as blazing star (*Liatris* sp.) may benefit from or require stratification for germination.

and allow them to dry at room temperature. Using a seedling heat mat, sow and germinate the seeds per the instructions on the seed packet. Dry stratification involves storing dry seeds at freezing temperatures for a month or more, depending on the species. Never put moistened seeds in the freezer, as this can kill them. Tropical seeds, such as tomato and pepper, do not appreciate any sort of chilling.

Seeds of many native wildflowers and prairie grasses will benefit from dry stratification for 30 to 90 days before you try to germinate them. Other native and woodland-type wildflowers perform best with moist stratification for about a month. Research the plant you want to germinate to determine whether it needs any chitting, scarification, or dry or moist stratification to germinate successfully.

Growing Media

When buying or mixing your own growing media (potting soil or soilless mix) for seedlings, keep in mind that you need to create a balance between water retention and good drainage. Small seedlings can dry out quickly and die, so they need growing media that can hold adequate moisture, but they are also susceptible to many soil-borne fungal diseases brought on by excess moisture.

The more organic matter in the growing media, the more opportunity for diseases and pathogens to become a problem. It is beneficial to start your seeds in a sterile, soilless growing media that contains peat moss, coir, or sphagnum moss mixed with vermiculite and horticultural perlite. You can even mix your own, using a 1:1:1 ratio of the three primary ingredients. This mixture creates a light, fluffy substrate that holds moisture but allows enough air space for the delicate roots of tiny seedlings.

Coir, or coco fiber, is a more sustainably harvested and renewable alternative to peat moss. Peat moss is harvested from living peat bogs, which are not replaceable, but peat bogs keep making more peat. If we don't harvest more peat than the bogs produce naturally, peat moss is sustainable. The pro-coir argument is that coconut trees can be grown anew. But some types of coconut are now considered endangered because of fungal and bacterial diseases that have killed many trees, so the availability, cost, and sustainability of coir may change.

There are many types of premixed potting soil available. Select a blend specifically appropriate for seeds, larger indoor plants, or outdoor plants, each of which has different characteristics.

Coir is excellent for growing seedlings and mixing into the growing media for larger plants. You can use coir alone or in combination with other ingredients, or even as a hydroponic substrate. It holds moisture well for an extended period while maintaining a porous structure that drains well. Some coir can be high in salts, which can interfere with the uptake of certain nutrients, such as calcium. If you plan to use coir as a hydroponic substrate, look for brands that have been washed or composted.

PROPAGATION AND PLANT CARE

Loose coconut fiber is an excellent soil amendment for aeration and moisture retention.

FAR RIGHT Coir mats can be used to propagate, grow microgreens, or line planters.

Many conventional soilless seed-starting mixes are precharged with a granular synthetic fertilizer. If you are concerned about consuming foods grown with chemicals and prefer more organic gardening practices, use a bioactive soil instead. Look for a lightweight, soil-based seed-starting mix that contains organic matter and is labeled organic. These mixes are often premixed with natural organic components that will be broken down as fertilizer over time (such as worm castings, seaweed, and humus). They are a bit heavier than the soilless mixes, and you may experience more variation between dryness and wetness. There is also an increased risk of soil-borne diseases.

Over time, as you experiment with different types of soils and potting mixes, you'll find your favorites. You may even prefer to create your own custom blend; there are many recipes available in books and online. Different plants do well in different types of mixes. As a general rule, when starting seedlings, use a lighter mix that holds a bit more moisture.

If you don't want to mix your own, look for packaged growing media labeled specifically for starting seeds. You can find plenty of conventional and sustainable options on the market. As your plants mature and grow larger, you will transition to mixes that can be heavier and provide better drainage.

RECIPE FOR SEED-GROWING MEDIA

4 parts fine-screened organic compost
2 parts coir, moistened
1 part perlite
1 part vermiculite

Growing Container

When choosing a container for starting your seeds, look for one that is only 2 to 3 inches deep and has drainage holes. Larger containers that hold more soil may cause problems for tiny seedlings, as they can hold too much moisture or wick moisture away from the small root system.

You can purchase plastic seed-starting trays that are segmented into small cells or plugs, small plastic pots, biodegradable plug packs, small clay pots, or any number of other options. If you're into recycling and like to use what you have on hand, try egg cartons, plastic containers from the grocery store, Styrofoam cups, and the like. If the container has no drainage hole, create one. If you try to grow plants in containers without a drainage hole, you will quickly drown them.

Before you make your selection, think about whether you want to reuse the container. A reusable container will be more stable and permanent, but biodegradable seed plugs allow you to plant the entire cell directly into a larger pot or your garden without lifting out the seedling or disturbing its root system.

Soilless seedling peat pellets, also known as seed plugs, are another easy option. These are compressed dry disks of peat, plus a small amount of fertilizer, wrapped in a biodegradable film. Place the pellets in a seedling tray and cover them with water, and the pellets will expand. Set the pellets in a solid seed tray (no need for an additional container), drop your seeds into the opening at the top of the film, and keep moist. Seed plugs are sold with small seed-germinating greenhouse kits.

If you're looking to grow organic and sustainable, consider making your own seed plugs. You can use a small newspaper pot maker to create the small degradable vessel, or try cheesecloth as a casing. Substitute coir (or a lightweight organic seedling mix) for peat, and you have your own green seed plugs.

A soil blocker is a metal form that creates compressed square soil plugs out of your growing media. These soil blockers sit in a solid seed tray, just like peat pellets or home-made seed plugs. It can be challenging to get lightweight soilless mixes to hold their shape, and they may require a binder in the form of denser organic matter

TOP There are many biodegradable container options for seed starting.

BOTTOM You can use small plastic reusable containers to start seeds and take cuttings.

129

BELOW AND RIGHT You can direct seed some seeds and root crops (such as lettuce and beets, respectively) into their final container. Manage the moisture well, and make sure the soil doesn't stay too soggy.

ABOVE Soak the seed plugs in water until they expand to their full size. Now they are ready for seeds.

LEFT Compressed seed plugs in a seed-starting tray.

or a bit of clay-based soil. Of course, reintroducing bioactive matter into your soilless mix means it is no longer sterile. If you're an organic gardener, this may be more desirable.

When you use pellets, seed plugs, or soil blocks, allow the seedling to grow in the small casing until the roots hit the edge of the netting or newspaper, then transplant it to a larger container.

If you are starting seeds in a hydroponic setup, you will likely use root plugs to start your seeds and grow your young plants. If you buy a self-contained hydroponic growing system, these root plugs may come with the unit, or you can buy them separately.

Root plugs are typically made from hydrophilic, or water-absorbent, polyurethane foam. Biodegradable organic root plugs are also available. Root plugs are often inoculated with beneficial microbes to help seeds and cuttings get off to a great start. You can use root plugs to start seedlings that remain in a hydroponic setup, or use organic root plugs to grow and transplant seedlings into soil-based growing media or the outdoor garden. You can also sprinkle your seeds directly into the hydroponic containers with a substrate such as hydroton or coir.

No matter what container you use, it's best to place the containers or plugs into a seed-starting tray without holes. A solid tray will catch water runoff from the containers and hold a bit of extra moisture as the seedling roots mature. You can also start microgreens in these trays.

TOP Basil cuttings in foam rooting plugs.

BOTTOM These expanded clay pellets, known as hydroton, are often used as a supportive substrate in hydroponic and aeroponic growing systems.

Sowing Your Seeds

Before planting seeds, be sure to thoroughly moisten the media, pellets, or plugs. Plant your seeds at the depth recommended on the packet. Different plants prefer different depths. Typically you can plant larger seeds deeper and smaller ones more shallowly. You should sow some seeds, such as lettuce, on top of the soil and leave uncovered.

You won't always achieve 100 percent seed germination, whether you buy or harvest and keep for the next season. Sow two seeds per container, cell, or plug, so if one seed doesn't germinate you have a backup.

Moisture

It is key to maintain proper moisture levels for your seedlings. The growing media should always be damp to the touch, similar to a wrung-out sponge. Never let it dry out,

Different types of seed will germinate faster than others. This basil is clearly an overachiever.

but don't let it stay soggy either. Do not overwater seeds that haven't yet germinated, as they can rot. Use a plant mister bottle to keep the soil surface moist until seeds germinate, then begin adding a little water to the seed tray so the growing media can absorb it as seedlings mature.

It is also important to increase humidity around your seeds as they are germinating. The easiest way to accomplish this is to place a humidity dome on top of your seed trays. Remove the dome once all your seedlings have germinated and begin to produce green growth. If you keep it on too long, fungal diseases or rot could set in. While young seedlings need to remain moist, excess moisture can promote damping off disease, which causes seedlings to rot at the soil line and topple over. Too much humidity also encourages powdery mildew on young seedlings and transplants.

Soggy seed trays can drown your small seedlings. Capillary mats keep adequate and consistent moisture at the root level for your seedlings without drowning them. Set these mats, which resemble felt, in the bottom of a solid seed tray, and place seed plugs or seed pots on top. Add just enough water to saturate the mat. Capillary mats are also handy for growing microgreens.

Temperature

The beauty of indoor gardening is that you can grow a bevy of food crops at the same time, regardless of outdoor weather, but you must mimic their basic environmental needs. If you have struggled to get warm-season crop seeds to germinate well, it may be that the growing media wasn't warm enough. Both air and soil temperature affect the speed and success rate of seed germination and growth.

A good humidity dome comes with a vent you can open and close to conserve or vent moisture and heat.

ABOVE Seeds germinating under humidity domes.

LEFT Microgreen seeds growing on a felt-like capillary mat set inside a solid plant tray.

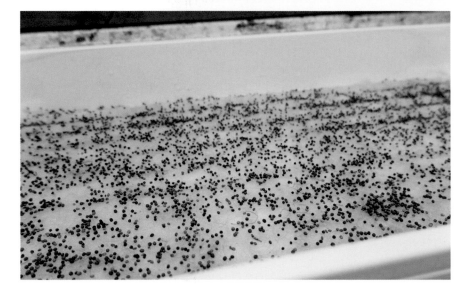

A heat mat warms up the soil for new seedlings.

If you are growing warm-season crops—such as basil, tomatoes, or peppers—from seed indoors during cold winter months, you must warm their soil. If you are growing cold-weather lovers such as spinach indoors in the summer, you have to cut back on the heat.

Each type of plant has a different optimal temperature range, based on its natural environment. Be sure to check the seed packet for specific optimal temperature ranges. Many seeds germinate well in the 68°F to 78°F (20–26°C) range for both soil and air temperature. If temperatures are too cold or too warm, some seeds take a very long time to germinate or may not germinate at all.

Ultimately, your goal is to get as many seeds to germinate as quickly as possible. For crops that need warmer soil temperatures, especially when you're starting them indoors during colder months, a seedling heat mat can speed up the germination process and ensure success. Heat mats, which are placed directly under your seed tray, can warm soil to an average of 15 to 20 degrees Fahrenheit above room temperature. Most homes have an average temperature in the 68°F to 75°F (20–24°C) range, but may be cooler in winter months.

Be sure to use a heat mat specifically designed for seed germination. Seedling heat mats are insulated and moisture-resistant, so you can place them on normal surfaces under your seed trays. (Do not try to repurpose a heating pad designed to soothe your aching back. This presents a fire hazard.) There are different sizes of heat mats, and you can even buy mats that link together to accommodate multiple seed trays. Once your seeds have germinated and begin growing, you can remove the heat mat. If your growing space is a bit cool, keep the heat mat in place as the plants mature.

If you want to keep close watch on temperature, add a thermostat to your heat mat so you can set a specific soil temperature for your seeds. The thermostat will also lower the temperature of the heat mat or turn it off if the heat in your home matches the desired set point. This prevents your seedlings from frying.

Your seed packet will offer estimated germination times based on optimal temperature ranges, but you will likely germinate seeds much faster in a controlled environment with heat mats. For example, I often have germination occur on lettuce seeds and mixed seeds of microgreens within only two to three days of seeding.

Culling the Herd

This part can be tough: you are going to have to murder some of your seedlings. Remember those two or three seeds you placed into the seed cell? Sometimes all of them germinate. Or you may have inadvertently dropped more than a few seeds into the plug cell. When most or all of the seeds germinate, you are confronted with a tiny seedling forest. It is tempting to let them all continue to grow, but your seedlings will be better off if you cull the weakest ones. More than one seedling per cell causes too much shading and resource competition, resulting in weaker seedlings overall. After your seeds have sprouted, choose the strongest, stockiest seedling in each cell and snip the remaining seedlings at the base. Throw the excess seedlings on your salad or feed them to any critters who would appreciate some greens.

If multiple seeds germinate in the same cell, keep the strongest and snip off the extras.

Seasonal Timing

When starting seedlings indoors for your outdoor garden, you must prepare the seeds early enough to plant outside at the optimal time. Crops such as broccoli, kale, and cabbage can be ready to plant outdoors in as few as 5 or 6 weeks from the date you bump them up to a 4-inch pot. Tomatoes, on the other hand, will need a good 8 to 10 weeks before transplants are ready to go outdoors. Leeks and celery can take a solid 12 weeks. Research how long it takes to grow a planting-ready transplant, and then count backward from the outdoor planting date for your area. That's when you have to get your seeds started indoors. This is also the amount of time you'll need to grow transplants large enough to plant into larger containers for long-term indoor growing.

Seed Depth	Seed Spacing	Soil Temp for Germ.	Days to Germination	Thin Plants to
¼"	See Below	70-90°F	8-25	12-18"

Sowing Indoors-Start seeds 8-10 weeks before your average last frost date. Use a sterile seedling mix and keep uniformly moist. Germination may be slow and erratic. Provide the seedlings with plenty of light to produce strong, high-yielding, mature plants. Use a heat mat to achieve the optimum soil temperature.
Sowing Outdoors-Not recommended.
Growing Tips-The number of early flowers can be increased by giving the plants a cold treatment before transplanting outside. This is done by exposing transplants to temperatures of 55°F during the night and 70°F in the day. Cold treatment should occur only after there are 3 sets of true leaves. Transplant out when soil temperature has warmed to 65°F.
Fertilization Tips-High phosphorous soil amendments such as bone meal will help get the plants off to a fast start. Apply ½ cup of our complete fertilizer around each plant to provide the nutrition necessary for optimum production.
Seed Specs-Min. germ. standard: 70%. Usual seed life: 2 years.
Please read our seed guarantee before opening this envelope.

Days-to-Harvest

On the seed packet you will find information on when you can expect flowers or fruit from a mature plant, also known as days-to-harvest. If your tomato seed packet states 85 days, that means it typically takes 85 days to start harvesting fruit. But 85 days from when, exactly?

The number of days-to-harvest can vary significantly, depending on whether you

FAR LEFT Refer to your seed packets for seed timing, depth, and spacing instructions.

direct seed a crop or grow a transplant. If you direct seed your crop into the garden or the container in which it will grow indoors, the days-to-harvest number is calculated from the date of germination. For crops that are first grown as a transplant that will be repotted, the days-to-harvest is calculated from the date of transplanting. You restart the days-to-harvest number once you establish the transplant in its permanent home.

For the 85-day tomato variety, harvest time comes in 85 days plus the 8 to 10 weeks it took to grow a garden-ready transplant from seed, or 135 to 149 days (19 to 21 weeks) to produce harvestable fruit. Some crops are much quicker than others. Radishes, for example, can be ready to harvest within 28 to 35 days of germination.

Root crops (carrots, turnips, beets, radishes) and large-seeded crops (beans, broad beans, squash, corn) do not transplant well, so you should sow them directly into the garden or final growing container. In this case, the days-to-harvest number on the seed packet is the time it will take to harvest from the date of germination. Plant these crops when indoor or outdoor temperatures are ideal. Tropical crops (eggplant, peppers, and tomatoes) are typically started early indoors as transplants, allowing for bigger plants in the outdoor garden once temperatures are appropriate or after plants are bumped up to larger containers indoors. For these plants, days-to-harvest is counted from the date they are planted into the final container or outdoor garden.

Other crops fall somewhere in the middle. You can direct seed lettuce and many other salad greens, for

TOP Days-to-harvest varies, depending on whether you direct seed or transplant your edible crops.

BOTTOM Root crops don't do well when transplanted. Seed them right into the garden or final growing container.

example, into their final container or garden spot, but you can also grow them as transplants first, then bump them up into larger containers or set them in an outside garden.

Collecting and Storing Seeds

Collecting and saving seeds from the plants you're already growing is a sustainable approach to your gardening endeavors and a smart way to save money. There are two main types of seeds: wet and dry. Wet seeds are produced from fleshy fruits such as tomatoes, eggplant, and squash. They typically remain inside a large amount of flesh and are not visible unless you break open the fruit. Dry seeds develop inside a husk or a fruit pod that dries completely so seeds are visible. Beans, okra, peppers, onions, and herbs such as dill produce dry seeds.

Harvest dry seeds while still on the plant, once they have completely matured and dried. The dry pods will be easy to open or will burst open on their own, and the seeds inside will be hard and completely dry. Dry seeds are easy to collect and store. Once you have harvested the seeds, leave them on a towel in a cool, dry place for a few days to ensure they are completely dry before you store them. Place them in a sealed container—such as an envelope, a paper seed packet, or a jar—and store in a cool, dry, dark place.

Harvesting and storing wet seeds is a little trickier. First, rinse thoroughly to separate the seeds from the fleshy part of the fruit. After the seeds are completely clean, spread them out on a dry surface for several days before storing.

Some wet seeds perform best if you ferment them before storing. Fermenting mimics the process the seeds go through on the vine as they mature and enables you to eliminate fungal activity that could damage the seeds or prevent germination in the future. Fermentation also breaks down germination inhibitors that may remain in the seed coat.

To ferment seeds, place them in a bowl of water with some of the fruit flesh remaining around them. Use about twice the volume of water as the amount of seed. Place the container in a warm area, where temperatures are between 75°F and 80°F (24–26°C), for two to five days. (Alternatively, look for bubbling or mold to appear on the surface of the water, which could occur sooner.) Once the fermentation is complete, the viable seeds will sink to the bottom of the container and the bad seeds and any remaining debris will float to the top. Do not let the seeds ferment too long, or they can start to germinate. Spread out the good seeds on a dry surface for several days, then store as you would dry seeds.

Keeping seeds in a cool, dry place will help extend their life. Be aware, however, that the longer you store seeds, the more their germination rate decreases. For example, you may achieve 100 percent germination from seeds you've harvested and stored within the same year. But three years from now you may get only 60 percent germination. These rates will differ depending on plant variety and the conditions in which seeds are stored.

TOP These butterfly weed seed (*Asclepias* sp.) are dry and ready to be harvested.

BOTTOM Once tomatoes are large enough to develop mature seed, you can separate them from the fruit flesh. Tomato seeds can be fermented prior to storing.

137

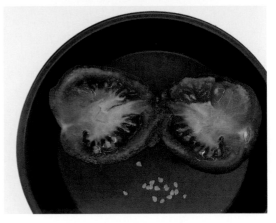

I use small, sealable envelopes to collect and store my dry seeds.

RIGHT Some plants, such as salvias, are very easy to propagate from stem cuttings.

RAISING PLANTS FROM CUTTINGS

When growing a mature plant from seed will take longer than you want to wait, seeds are not available for the type of plant you want to grow, or you want to reproduce a hybrid, you must grow from plant cuttings. Vegetative reproduction is often referred to as cloning. Taking cuttings from a healthy plant can provide you with a continual supply of new, identical plants. While not all plants root well from cuttings, many do.

It can take anywhere from 8 to 12 weeks to grow a garden-ready or big container-ready tomato transplant from seed. But you can root a vegetative cutting from an existing tomato plant in about a quarter of that time.

Mother Plants

The overall health and nutrient levels of the mother plant will have a big impact on the success of your cuttings. Be sure to start with a vigorous, healthy mother plant that has characteristics you want to replicate. Cuttings tend to root better and take off faster if the mother plant has high levels of carbohydrates and less nitrogen. New cuttings need to establish a good root system before pushing out new leafy growth they cannot yet support. Stop fertilizing your mother plant with nitrogen about a week before you take cuttings.

Where to Take Cuttings

Plants have varying requirements for how, where, and when to take cuttings, the ideal size of the cuttings, and how long they may take to root. Some plants, such as succulents, will root from the base of a leaf. Others, like begonias and African violets, will root straight from the leaf surface. Many perennials are propagated by taking root cuttings or divisions. You can propagate certain tropicals, such as

LEFT TO RIGHT

Rose plant stem cuttings.

African violet leaf cutting.

Baby African violets from leaf stem cuttings.

LEFT TO RIGHT

This begonia leaf has been damaged, which caused it to develop tiny baby plantlets along the leaf vein.

A begonia leaf cutting sits flat on the soil surface.

Most types of strawberries produce runners, or elongated stems with small plantlets that can be individually rooted.

LEFT TO RIGHT

Many tropicals, perennials, annuals, and edibles root directly in water. Transplant to soil when plants develop a small root mat.

A rooted plantlet developing at the base of a succulent leaf.

Some plants develop lateral or aerial roots that show where you can root them. This section of tomato stem begs to be rooted.

pothos ivy, by setting a stem in water for a few weeks. Remove some of the existing foliage from the cutting to reduce excess transpiration. Some plants cannot be propagated from vegetative cuttings, but you can divide them at the root zone.

Rooting Hormones
The trick with taking cuttings, just as with germinating seeds, is to get the cutting to root before the tissue rots or dies. The sooner the rooting occurs, the more likely the cutting will grow successfully. You can speed up the process by using supplemental rooting hormones, which are chemicals found naturally in plant tissues. Apply these to your cutting to encourage faster and more vigorous root development. Rooting hormones are available in gel or powder form. Dip the base of your cutting into the rooting hormone, making sure to coat the entire tip, and place it in your rooting media or substrate.

Seaweed extract also contains trace amounts of rooting hormones, plus additional important nutrients. You can use it as a natural rooting hormone and to reduce transplant shock.

Soak or dip stem cuttings in rooting hormone before sticking. Always follow label instructions.

Rooting Substrate
While you can root your cuttings directly in potting soil, you can also use inert media. Bioactive soils may contain pathogens that can rot the stem of your cutting before it has a chance to produce roots. If you use potting soil, be sure not to let it stay soggy during the rooting process.

To avoid soil pathogen problems, use media such as coir, rockwool, Oasis cubes, or anything made with perlite and peat. These help balance the moisture-to-air ratio available to the cutting and reduce pathogens. Presoak the rooting media with dechlorinated or distilled water. You can naturally dechlorinate water by leaving it in a lit area for 24 hours. However, your municipal water may contain other chlorinating products that will take longer to break down. You can add a dechlorinator to your water to speed up the process.

Lighting Your Cuttings
While young seedlings typically need high light levels to thrive, intense light can burn vegetative cuttings. Typically, ambient light from a bright window is strong enough for unrooted cuttings and those that have not yet put on any new growth.

You can expose cuttings to 24 hours of light to speed up the rooting process. Cool-spectrum T5 fluorescents or CFLs are good options for rooting vegetative cuttings, but be sure to place the lights at least 2 feet above the cuttings, if not higher. If you place the lights too low, you can burn the cuttings. Leave the lights on for 24 hours a day until the cuttings have roots. Once the cuttings have rooted

Plants that are fast rooters, such as angel's trumpet (*Brugmansia* sp.), usually do just fine rooted directly into potting soil. They root before they rot.

STEP-BY-STEP INSTRUCTIONS FOR TAKING CUTTINGS

SUPPLIES
- mother plant
- sharp snips
- rooting hormone
- growing media
- propagation tray
- humidity dome
- heat mat

1. Disinfect your snips using alcohol or a 10 percent bleach solution. Make a clean cut about 4 inches below a relatively new stem that has no flowers or buds on it. Take a 2- to 4-inch cutting. Remove any additional side stems or leaves at the bottom of the cutting, leaving about 1 inch of bare stem. Be sure that the cutting includes at least one leaf node, which is where the leaves emerge from the main stem and where new root tissue will form. Be careful not to damage the stem base.

2. Dip the bare end of the stem into the rooting hormone.

3. Gently place or stick the cutting into the rooting media. If your humidity dome has a vent, open it.

4. Cover all cuttings with a humidity dome. Don't overcrowd: while you have to maintain humidity around the cuttings during rooting, they also need good air circulation.

5. Keep the rooting media moist, but not soggy.

6. Site cuttings away from intense or high levels of light, which could burn them. Bright light from a window or a couple of feet away from fluorescent or CFL fixtures will be sufficient until the cuttings have rooted.

7. Give it time. Some plants can take a couple of months to root, while others will root in a week's time. For plants that like warm growing conditions, or if you're starting cuttings in winter or in a cool part of your home, use a seedling heat mat under the rooting tray to speed things up.

and start pushing out new growth, you can place them under brighter lighting suitable for that type of plant and on the correct day-night lighting duration.

Propagation Incubators

Propagation incubators can make cloning much easier. They typically use hydroponic or aeroponic methods for root cuttings. These setups will provide moisture, air, and rooting hormones at the right levels to deliver a high rate of success with your cuttings.

There are several other methods for vegetative propagation, such as air layering, grafting, and tissue culture. But many wonderful books go into much more detail on these techniques, so I will leave you to investigate elsewhere.

FAR LEFT A deep-water propagator.

An aeroponic propagator.

Roots developing on a salvia cutting.

FAR LEFT Cuttings of salvia, tomatoes, citrus, and more in the propagator.

PROPAGATION AND PLANT CARE

A small lettuce in a seed plug is rooted and ready to bump up into a 4-inch pot. After the plant roots fill the new pot, you can move it to a 1-gallon container for its long-term home.

TRANSPLANTING YOUR SEEDLINGS AND CUTTINGS

A common newbie mistake is to allow the seedlings or cuttings to remain in their original seed plug or cell for too long. Those tiny 2-inch pots and pods are not meant to sustain your plant indefinitely. Once the roots of your seedlings or cuttings have grown to the edges of their seed plug or pot and have sprouted three or more true leaves, it is time to bump them up to larger containers. True leaves look like mature leaves. Once they emerge, the process of photosynthesis has begun and the seedling is no longer relying on the resources stored in the seed. When your rooted cutting does the same, it is ready for a bigger home.

Don't overdo it when selecting the size of your new container. Replant a 2-inch seedling plug or cutting into a 4-inch diameter pot. If you plant a very small seedling in a much bigger container, it may drown in too much water held by larger soil volume or it could dry out as water moves away from the root ball to the edges of the container.

Always choose growing containers that have drainage holes in the bottom, or porous fabric grow pots that allow water and air to permeate. No drainage holes equals drowned plants. Use water-catching trays or tubs to collect the excess water that drains from your pots.

Fill your transplant pot with a quality potting soil, then gently transfer your seedling or cutting into the new pot at the same depth at which the seedling naturally emerges. If you're using a plug tray filled with a growing media, push your finger up against the bottom of the plug (or partially through the drainage hole) to loosen and slide the seedling or cutting out of the tray without damaging it. You can also use a knife or spoon. Never pull on a seedling's stem. If the seedling has a good root system, you may be able to squeeze both sides of the plug and pop it right out. If there is not much of a root system on the seedling, it is probably too early to transplant. Hold the seedling transplant underneath the root ball to set it into its new container.

Now that your transplants are actively photosynthesizing and growing, they will take up water faster and can dry out quickly under grow lights. Check their water needs more frequently than when the tiny seedlings were safe under a humidity dome.

Once the root system has filled in to reach the edges and bottom of its new pot and is extensive enough to hold together most of the soil in the pot, your transplant is ready to be planted into your outdoor garden (if temperatures are right) or moved into a larger container for continued growth indoors. This can take anywhere from 5 to 10 weeks for seedlings, faster for rooted cuttings. The type of crop and the growth rate will dictate what size pot to choose. Again, don't go too big too fast. Your transplant may require several size upgrades before it is large enough to be planted into its final container.

FEEDING YOUR YOUNG PLANTS

As your seedlings or cuttings grow, they will need more than just water and light to survive.

Feeding Seedlings

Once your seedlings or cuttings have developed four or five true leaves, it is time to start feeding them with fertilizer to make sure they have the nutrients they need to thrive. Small seedlings cannot handle full-strength fertilizer, so dilute your product with water by one quarter to one half the recommended rate.

BELOW, LEFT TO RIGHT

This tomato transplant is filling out its 4-inch container.

Now that the tomato transplant has rooted out to all sides of the 4-inch pot, it is ready to be bumped up into a larger container or planted into the outdoor garden.

Now that the tomato transplant is settled in its new home, you can start your days-to-harvest clock.

PROPAGATION AND PLANT CARE

Chlorosis can appear as an overall yellow cast to the leaf or interveinal chlorosis (yellow between green veins), as shown, because of a lack of chlorophyll. Chlorosis signals potential nutrient deficiencies (iron, manganese, zinc); it can also signal poor drainage, soil compaction, or high soil or water pH.

You can use many fertilizers to get your seedlings and cuttings off to a great start, but it's best to use a liquid suspension rather than a granular formula. Liquid seaweed, fish emulsion, and liquid humus products are good natural options that won't burn your young plants. Dilute these products to half strength when applying to small seedlings, or choose a liquid fertilizer labeled specifically for seedlings and follow the manufacturer's instructions. If you choose a synthetic fertilizer, dilute it by one quarter of full strength, as these products are more likely to burn or damage your young plants than naturally derived fertilizers.

Feeding your seedlings once per week at low doses is adequate, or every two weeks if you have bumped up to a stronger feed. With cuttings, wait until they have started to develop roots. As plants mature you can increase the rate of fertilizer application to full strength. Do not mix your fertilizer at a stronger-than-recommended rate. While there is less danger of damaging your seedlings with too much organic fertilizer (you'll just be wasting product), if you overdose with a synthetic fertilizer you could burn your plants beyond repair.

TOP These plants are beginning to set fruit. As more fruit begins to develop, I will start fertilizing plants again.

BOTTOM When plants are fruiting and ripening, be sure to keep them on a regular fertilizer regimen.

Feeding Larger Plants

Once vegetable crops or blooming plants mature and begin to flower, dial back the nitrogen fertilizers. Too much nitrogen encourages plants to put their energy into leafy green growth instead of flowers and fruit. Once plants enter flowering mode, you can switch to a low-nitrogen fertilizer or hold off altogether until baby buds and fruits begin to appear. Once vegetable crops start to set fruit and fruit reaches about half its mature size, pick up your normal every-other-week fertilizer rotation. The plant will now put those nutrients to work developing flowers or fruits. Each type of plant will have its own needs, so research your specific crop to choose the best fertilizer.

Other excellent solid fertilizer and soil amendments include worm castings, kelp meal, greensand, lava sand, blood meal, bone meal, fish meal, crushed crab shells, dry molasses, and more. You can also mix your own custom natural fertilizers using a combination of ingredients.

I grow most of my plants naturally and organically, so I stick with bioactive soils, naturally derived fertilizers, and low-impact pesticides, especially for

edible crops. Occasionally I resort to a synthetic product. When using synthetic formulas, take care and do not overapply. Also, be mindful of where you are using these products and know the potential for environmental impact.

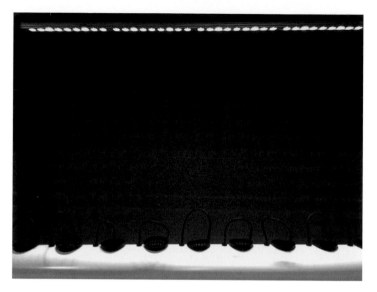

A hydroponic grow system with baskets ready for young plants.

HYDROPONICS

Many indoor growers choose to skip the soil and grow their plants in hydroponic, aeroponic, or aquaponic systems. While growing hydroponically will reduce some of the pressure of checking your plants daily for water needs, you will have to monitor the system.

Hydroponic systems use liquid, gravel, sand, or other soilless medium with added nutrients to grow plants. You can grow most plants hydroponically, although there are some limitations. The most common form of hydroponic system is the nutrient film technique (NFT), which involves growing plants with their roots directly suspended in water. Other methods include wick, water culture, ebb and flow, drip, and aeroponics.

Most hydroponic systems combine a few simple parts: the container, a water reservoir that holds the water, a pump that moves the water around the plant's roots, and an air pump to keep the water oxygenated. You add nutrients and products to control disease and water pH.

Aeroponic systems use water to mist plant roots several times a day. Plants are suspended in small nets or baskets that may contain clay pellets or perlite to provide stability. By misting plant roots intermittently instead of keeping the roots suspended under water, you can cut back on potential issues with oxygenation or disease problems.

Aquaponic systems combine hydroponic growing and aquaculture, or fish farming. Plant roots are suspended in a grow bed. Fish and other aquatic animals inhabit a tank with a gravel bed at the bottom. The water from the tank and the waste produced by the fish are circulated into the grow bed to feed the plants. The plants then filter the water, which is circulated back to the fish tank.

A large grow tent set up with an ebb and flow hydroponic bucket system for multiple large plants.

Herbs growing in an aquaponic system. Notice the aquarium beneath the fixture.

Some aquaponic gardens are very large and must be sited outdoors or in a garage, but you can place a mini aquaponic garden on your counter. Don't forget the grow lamp.

Hydroponic systems require special fertilizer and water-management products to keep the system in balance. You must test your water regularly, and systems need consistent cleaning and maintenance. Aquaponic systems also add the complication of maintaining animal life. They are fascinating, and excellent for growing plants indoors.

INDOOR POLLINATION

One very important thing to remember about growing fruits indoors: the flowers require pollination to develop fruit. Wind and pollinating insects would accomplish that work outdoors, but indoor gardeners must play the role of honeybee. Some plants are more self-sufficient when it comes to pollination. Tomatoes, for example, have both female and male reproductive parts within the same flower—they just need a little jostling to move the pollen around. Air circulation from a fan is usually adequate, but I like to shake my tomato plants now and then once they start flowering to give them a little boost.

Other plants require more help. Cucumbers, for example, produce separate female and male flowers. You must manually move the pollen from the male flowers to the female flowers, which could be a good distance away on the vine. Flowers open in the morning, and pollen is viable only that day. You must use a small paintbrush to pick up some pollen from a fresh male flower and brush it onto the pistil of an open female flower.

PLANTS

EDIBLE PLANTS

This chapter includes a core list of edibles that are well suited to growing under lights indoors. I have omitted certain fruit and vegetable crops because their vernalization or overall temperature requirements are too difficult to replicate in a typical indoor environment. Don't be afraid to try the more challenging plants—experimentation is how you learn.

GROUP PLANTS BY LIGHT

It is important to group together crops with similar lighting and temperature requirements. If you mix plants with different environmental needs, one will likely thrive while the other fails. For example, if you grow peas together with your lettuce under 10 to 12 hours of light, the lettuce will probably perform as expected, but the peas will never flower or fruit. While both lettuce and peas are long-day plants, you want to discourage flowering in lettuce but encourage it in peas. This means that while you'll keep your lettuce under a shorter daylength to prevent flowering, you'll have to move your peas to light that stays on longer than 12 hours to achieve flowering and fruit development.

GROUP BY TEMPERATURE

Grouping by temperature preferences is also important. Remember that cool-season crops, such as lettuce, greens, and peas, may not withstand the heat build-up inside an enclosed grow space or grow tent unless it is artificially cooled. However, natural room temperatures in open spaces in your home, basement, or garage, combined with cooler running lamps, can be good options. Your heat-lovers, such as tomatoes, peppers, and cannabis, can handle tighter spaces with warmer HID lighting.

SIZE MATTERS

When choosing varieties of edibles to grow indoors, look for dwarf, bush, container, or patio varieties. These are compact versions of taller or vining edibles that

TOP LEFT Lunch from the indoor garden.

MIDDLE LEFT Grow dill with short photoperiods to suppress flowering.

BOTTOM LEFT Pepper plants need lots of light and warmth to produce plenty of fruit.

'Floral Spires Lavender' basil flowers faster under long photoperiods.

You must train indeterminate vining types of tomatoes onto support cables or trellises that will support their ongoing growth and increased size.

If you grow vining tomatoes, be prepared for a tomato forest.

are easier to accommodate in an indoor environment or in containers. While a standard squash plant can send out vines 15 feet long, a bush-type squash will grow only 2 or 3 feet tall and wide, the perfect size for an indoor container. Beans and cucumbers are also available in bush varieties. Standard okra plants can tower over you, but dwarf varieties grow to only 2 to 3 feet tall. Most pepper plants are fairly compact and well suited to indoor container production. Remember that determinate tomatoes typically grow to about 4 feet tall and 2 feet wide, while indeterminate tomatoes are vines that can grow to anywhere from 6 to 25 feet tall.

If you intend to grow larger plants indoors over the long term, follow the recommendations for spacing and container sizes in each plant profile. You can maintain 4-inch containers of herbs under lights in your kitchen or anywhere in your home, but most root systems will outgrow this container and use up soil nutrients more quickly. If you harvest on plants in small containers regularly, they will eventually expire and you will have to replace them. Compost the old plant, or pot it up into a larger container and start fresh with a new plant in the small pot.

The following plant profiles provide you a general classification of photoperiod and best daylengths for each crop, temperature recommendations, and space needs. You can make necessary adjustments based on your own results.

Arugula, Rocket

Eruca sativa, Diplotaxis spp.

Arugula is a tough, prolific, easy-to-grow fresh green with the flavor of black pepper.

PHOTOPERIOD Long-day. Grow with 10 to 12 hours of light to suppress flowering.

LIGHT REQUIREMENTS Medium. Outdoors plants can produce well in partial shade. Cool-spectrum HO T5, CFL, LED are best. MH/LEC with air cooling. DLI 12 to 16 mol m^{-2} d^{-1}.

PROPAGATION Direct seed into garden or final containers indoors, or grow transplants for the garden or to pot into larger containers. Sow seeds ¼ inch deep at soil temperatures of 40°F to 55°F (4–13°C).

TEMPERATURE Plants grow best between 50°F and 65°F (10–18°C), but often tolerate and continue producing foliage at warmer temperatures. Normal indoor household temperatures are generally adequate.

SEASON Cool-season. Direct seed outdoors during the cool seasons. Transplants grown indoors can be planted during cool seasons outdoors, or grown indoors when outdoor temperatures are too warm.

SPACE Plants are compact with small root systems and are easy to grow in containers. Provide 6 inches of space between the base of each plant. Grow in 6- to 8-inch diameter or 2-gallon containers.

GROWING MEDIUM A variety in a pH range of 6.0 to 7.0. Use a loose mix with organic matter, or grow in hydroponic, aeroponic, or aquaponic systems.

WATER Moderate needs, but water consistently and evenly. Plants can thrive with less water at cooler temperatures.

FERTILIZATION Arugula is not a heavy feeder. Apply composted manure or a low N-P-K granular fertilizer at planting. If you harvest regularly, feed plants with a half-strength solution of liquid humus, seaweed, or fish emulsion once per month.

PESTS & DISEASES There aren't many pest problems, but you may experience aphids, whiteflies, fungus gnats, downy mildew, or bacterial leaf spot.

VARIETIES & CULTIVARS 'Astro', 'Esmee', 'Sylvetta', 'Wild Rocket'

HARVEST Plants grow quickly, and you can start harvesting leaves in 20 to 27 days from germination. Start picking when leaves are 2 to 3 inches long. The flowers are also edible. Harvest stems from outside the central growing point, which allows new leaves to emerge from the center.

Basil

Ocimum basilicum, O. americanum, O. sanctum

Basil is a heat-loving aromatic and medicinal herb that's a must-have in the kitchen year-round.

PHOTOPERIOD Facultative long-day. While the goal is to produce more leaves and fewer flowers on your plants, basil is stubborn when it comes to growing well under short daylengths, so these conditions will not improve your yields. Keep light periods in the 14- to 18-hour range and be prepared to pinch flowers.

LIGHT REQUIREMENTS Bright light indoors, with direct full-sun exposure for 6 to 8 hours outdoors. Choose cool-spectrum HO T5, CFL, LED, MH/CMH lamps.

PROPAGATION Direct seed, seeded transplants, cuttings. Seed germination rates are best at temperatures between 75°F and 85°F (24–29°C). For transplants, place two to three seeds in each cell and lightly cover with soil. Vegetative cuttings root quickly.

TEMPERATURE Plants need warm temperatures in the 75°F to 85°F (24–29°C) range, but also thrive in hotter temperatures. Plants are damaged or killed at temperatures below 50°F (10°C).

SEASON Can grow indoors year-round. Place seeds and transplants outdoors in late spring or summer.

SPACE Basil does not require a lot of space and grows easily in containers. Some varieties are very compact, reaching only 8 inches tall, while others can get much larger. If you have limited space, choose dwarf cultivars. Grow in 6- to 12-inch diameter or 1- to 5-gallon containers.

GROWING MEDIUM Loose, well-draining, with a pH range of 5.5 to 7.5. Plants are suited to hydroponic, aquaponic, and aeroponic systems.

WATER Moderate to light needs, but apply consistently and evenly, as plants have a shallow root system.

FERTILIZATION Plants are light feeders and typically do not require regular fertilization. Add a composted manure or low N-P-K fertilizer to the growing media at planting time.

PESTS & DISEASES Aphids, whiteflies, scale, powdery mildew, leaf blight.

VARIETIES & CULTIVARS 'Genovese', 'Boxwood', 'Dark Opal', 'Mammoth', 'Lettuce Leaf', holy, lemon, Thai

HARVEST Snip shoots as you need them. Store in a glass of water on your counter—never in the refrigerator.

Beans

Phaseolus spp.

Beans are nutritious, easy-to-grow legumes with varieties in many colors, shapes, and sizes. Stick to compact bush bean varieties for indoor growing, as they will reach only 2 to 3 feet tall, whereas pole beans can soar to 15 to 20 feet.

PHOTOPERIOD Varies. Some forms of common bean are day-neutral, while others are short-day plants, such as lima bean, soybean, mung bean, and yard-long bean. Grow short-day beans with a 12-hour photoperiod. Grow day-neutral beans with 14- to 16-hour photoperiods. If plants don't flower, try shortening the light period.

LIGHT REQUIREMENTS Bright light indoors, with direct full-sun exposure for 6 to 8 hours outdoors. Grow seedlings and small plants under cool-spectrum and switch to warm-spectrum for flowering, or grow under a blend. Use HO T5, CFL, LED, MH/CMH, or HPS lamps.

PROPAGATION Direct seed only. When growing indoors, seed your beans straight into their final container. Bean seedlings do not transplant well. Seeds germinate best at temperatures between 70°F and 80°F (21–26°C). At cooler soil temperatures, seeds will be slow to germinate and may rot.

TEMPERATURE After germination, beans grow and fruit best at temperatures between 70°F and 85°F (21–29°C). Plants can tolerate warmer day temperatures if night temperatures cool down.

SEASON Warm-season. Direct seed outdoors after possibility of frost has passed and soil temperatures have warmed. Grow indoors year-round.

SPACE Medium. Dwarf or bush beans will typically grow 2 to 3 feet tall and 1 to 2 feet wide, but can be planted densely and trained upright. Use 6- to 10-inch diameter or 2- to 5-gallon containers. You can plant multiple seeds in the larger container.

GROWING MEDIUM Loose, well-draining, rich with organic matter. Add coir to loosen the mix. Plants are suited to hydroponic, aquaponic, and aeroponic systems.

WATER Beans grow quickly and need consistent watering throughout the growing cycle. Don't let plants dry too much between watering, but they don't tolerate soggy soil or overwatering.

FERTILIZATION Apply a balanced vegetable fertilizer with a 1:2:2 ratio to the soil at planting time, then wait until plants begin to flower to fertilize a second time. You can use a diluted liquid fertilizer more frequently after flowering. Don't overfertilize with nitrogen or you will reduce fruit production.

PESTS & DISEASES Pest issues are not typical on beans grown indoors, but fungus gnats can be a problem and whiteflies could be an issue if you have an outbreak on other plants. Spider mites can appear on stressed plants. Powdery mildew and other fungal leaf spots can be a problem if you get water on the foliage or humidity is too high.

VARIETIES & CULTIVARS 'Bush Green', 'Bush Yellow', 'Dwarf Velour French', 'Carson', 'Bush Kentucky Wonder', 'Blue Lake 274', 'Contender'

HARVEST Be sure to harvest the pods before the seeds inside begin to mature. If your bean pod looks very bumpy, you already have bean seeds on the way and the pods are no longer edible.

GROW THE SAME WAY Black-eyed pea, lima bean, cowpea, pigeon pea, soybean, winged bean, yard-long bean

TIPS Legumes form beneficial relationships with certain soil bacteria to take up nitrogen from the soil. To add these bacteria to soil, look for packets labeled *Garden Inoculant*. You can also pretreat bean seeds with inoculant before you sow them, which can significantly boost yields.

Beet Root, Beets

Beta vulgaris

Beets are root vegetables that also offer edible leaves. There are many varieties of beets with a number of root colors, sizes, and flavors. Beets have a sweet flavor, and small or medium beets are generally more tender.

PHOTOPERIOD Long-day plant, but most beets also require vernalization to flower. Grow with a 12- to 14-hour photoperiod.

LIGHT REQUIREMENTS Bright to moderate light indoors, and can tolerate some shade outdoors. Better flavor is achieved with bright light. Cool-spectrum HO T5 fluorescent, CFL, and LED are best. MH/CMH lamps if air temperature is cooled.

PROPAGATION Direct seed. Seedlings do not transplant well. Sow seeds in final container at depth of ½ to 1 inch. Once seedlings emerge, thin their numbers to allow 1 to 3 inches between remaining seedlings. If you want bigger beets, thin the group even more to allow additional room for roots to expand.

TEMPERATURE Beets like it cool, ideally between 60°F and 65°F (15–18°C), but will grow at warmer temperatures and perform well. Good indoor spots are the

garage (insulated, if you're in a cold climate), the basement, and a glassed-in porch that stays on the cool side.

SEASON Cool-season. Direct seed outdoors during the cool seasons, or grow indoors when outdoor temperatures are too warm.

SPACE Direct seed into final container. Mature beets need 1 to 3 inches of space between each root, so choose wide, shallow containers.

GROWING MEDIUM Loose, well-draining, rich in organic matter.

WATER Apply consistently, but don't overwater. Beets can tolerate drying between watering, and often hold on to more nutrients when allowed to dry a bit. But don't be too stingy; beet roots can become tough and cracked if they don't get enough water.

FERTILIZATION Use a balanced granular vegetable fertilizer in the soil at seeding time. Then side-dress the beets or use a liquid fertilizer one or two more times during their growing cycle.

PESTS & DISEASES Aphids, damping off, leaf spot, downy mildew.

VARIETIES & CULTIVARS Many. Standards include 'Chioggia', 'Red Ace', 'Detroit Dark Red'. To save space and seed more densely, try miniature varieties such as 'Baby Beet' or 'Little Mini Ball'. 'Cylindra' forms long, narrow roots. If you just want tops for salads or microgreens, try 'Bull's Blood', which can be grown in low-light conditions.

HARVEST Harvest standard beets when they reach tennis-ball size, and mini beets when they resemble golf balls.

GROW THE SAME WAY Turnip, radish

TIP Beets, turnips, and radishes are quick-turn crops. Sow successions of new seeds every two to three weeks to keep a steady harvest coming.

Calabrese, Broccoli, Sprouting Broccoli, Rapini, Chinese Broccoli

Brassica oleracea var. *italica, B. oleracea* var. *alboglabra, B. rapa*

The name *broccoli* refers to the sprouting types that make smaller florets on longer stems, such as broccoli raab or rapini. Most of the large-headed broccoli you find at the grocery store is Calabrese. While I recommend you start seedlings of both types indoors for transplanting outdoors in the late summer and fall garden, they can be tough to grow to harvest indoors because they either benefit from (Calabrese) or require (sprouting broccoli) vernalization before they will flower. If you want to grow sprouting broccoli that can form flower buds indoors without cold temperatures, look for summer sprouting types. These don't require vernalization, but you still have to keep them at cool temperatures indoors.

PHOTOPERIOD Day-neutral. Grow with a 12- to 14-hour photoperiod.

LIGHT REQUIREMENTS Medium to bright. In the outdoor garden plants can tolerate a bit of shade. Use HO T5, CFL, or LED indoors to minimize heat.

PROPAGATION Seed. Sow seeds ¼ inch below soil level at soil temperatures between 70°F and 75°F (21–24°C). Seed indoors five to seven weeks before transplanting outdoors.

TEMPERATURE Calabrese and broccoli are cool-season crops, and temperature controls flowering responses; most do not handle heat well. Flower heads can also be malformed or damaged if young transplants sit at temperatures below 40°F (4°C) for a few weeks. The ideal temperature range for happy broccoli is between 65°F and 80°F (18–26°C). This can be tough to maintain in an enclosed grow room or grow tent unless it is air-conditioned, but normal household temperatures might be adequate. A cool basement or garage would be ideal.

SEASON Cool-season. You can direct seed outdoors in late summer and early fall in warm climates, then again two weeks before the last frost date. Plant indoors any time you can maintain optimal temperatures.

SPACE Calabrese and broccoli have shallow root systems, so you won't necessarily need large containers; 1- to 2-gallon pots will suffice. Calabrese foliage can spread wide, so be prepared to space out plants as they mature.

GROWING MEDIUM Well-draining, rich in organic matter. Add coir to keep soil moisture consistent. Plants are suited to hydroponic or aeroponic systems and prefer a pH range of 6.0 to 7.0.

WATER Keep plants evenly moist but not soggy, especially once flower heads begin to develop.

FERTILIZATION To form good heads, broccoli needs regular fertilization. Use a balanced vegetable fertilizer every other week. Side-dress plants with a granular fertilizer or water with a liquid fertilizer.

PESTS & DISEASES While cabbage loopers are one of the biggest outdoor pests, you probably won't have that problem indoors, unless you buy transplants that contain the eggs. Whiteflies and aphids can also be an issue.

VARIETIES & CULTIVARS 'Cruiser', 'Green Comet', 'Green Goliath', 'Premium Crop', 'Packman', 'Summer Purple', 'Romanesco Italia', rapini

HARVEST Cut Calabrese heads whole, at the base of the main stem, when a full head is formed and flower buds are still tight. You might get some side sprouts from the base of the plant after harvesting the main head. You can harvest the small florets of sprouting broccoli continuously through the growing season as they emerge.

GROW THE SAME WAY Brussels sprouts, cauliflower

CANNABIS, MARIJUANA

Cannabis sativa, C. indica

Midsize
cannabis plants.

Cannabis grows wild in both temperate and tropical areas around the world. It is used as a medicinal and recreational herb for its active compound, THC (delta-9 tetrahydro-cannabinol), and other compounds generally referred to as canna-binoids. Cannabis has been intensively selected, bred, and cultivated, and many varieties and cultivars vary widely from the original species and naturally occurring ecotypes. There are several main forms: marijuana, which is produced from the dried leaves and flowers and is typically smoked; kief and hashish, which are made by harvesting the trichomes that hold resin and then dried into blocks (which can be smoked or incorporated into edibles); and hash oil, a potent oil extracted from hashish. What we refer to as industrial or agricultural hemp are strains of *Cannabis sativa* that contain less than 0.3 percent concentration of THC (a designation that varies by state) that are grown for textiles, insulation, biofuel, clothing fabrics, and CBD extraction. If you intend to grow cannabis, make sure you clearly understand and abide by all federal, state, and local laws and restrictions.

PHOTOPERIOD Obligate short-day. The critical photoperiod for female cannabis plants is about 12½ hours, with about 13 hours for male plants. A traditional lighting regimen to bring plants into flower is 12 hours of dark followed by 12 hours of light.

The dark period must not be interrupted with any light, and it typically takes two weeks under this photoperiod regimen for plants to bud. However, the critical daylength will vary by variety or cultivar (typically referred to as *strain* in cannabis culture), and many growers use photoperiods ranging from 11 to 13 hours of light followed by 13 to 11 hours of uninterrupted darkness. You will need an enclosed growing space such as a grow tent, a sealed closet space, or a grow room. Some day-neutral strands.

Cannabis grows in two phases. During the vegetative phase you light plants for 18 to 24 hours to inhibit flowering and put on vegetative growth. If you don't have a lot of space and cannot accommodate large plants, you'll veg your plants for three to four weeks, then shift them to a flowering phase with alternating 12-hour cycles of light and dark (or the specific critical daylength required by your strain) to induce flowering. If you have more space and can accommodate larger plants, you can veg them for two to three months before inducing flowering.

There are also day-neutral strains of cannabis that are called auto-flowering. These plants are hybrids of *Cannabis sativa* or *C. indica* with *C. ruderalis*, a compact Russian species that evolved in areas with very long summer days. It flowers because of age and size rather than photoperiod. Auto-flowering hybrids can begin blooming as quickly as two to four weeks after sprouting, but overall have a shorter life cycle. These types tend to be more compact and can be grown under longer photoperiods for higher yields.

LIGHT REQUIREMENTS Cannabis is a sun-loving plant and needs 8 or more hours of direct sunlight outdoors, and high levels of intense light indoors. Newer strains reportedly tolerate much higher light levels, and amateur and professional cannabis growers use many different lighting strategies and setups.

TOP Cannabis flower buds.

BOTTOM Young cannabis clones.

Typically, HID lighting is considered the smart choice for efficiency and high quality and yields of cannabis. Vegetative plants are commonly grown under cool-spectrum MH or CMH lamps, then switched to warm-spectrum HPS lamps for flowering. If you have enough space in your grow tent or grow room, you can hang together both types of lamps, then switch from one to the other when shifting growing phases. Alternately, you may have a vegetative tent and a flowering tent with separate lights. However, many growers have good success by supplementing their HPS light with some additional blue light, using cool-spectrum HO T5 or LEDs during the flowering

Cannabis crop
growing under LEDs.

phase. Warmer-spectrum CMHs can also be a good option, as well as larger LED
rigs. Some growers have had good results with Dual Arc lamps (hybrid MH/HPS),
but others insist there is not enough light produced in each light spectrum when
combined, and thus yields may be lower.

If you are growing just a few plants in a very small space, CFLs produce a good
spectrum for cannabis and can be placed much closer to your plants than HID
lamps. HO T5 fluorescents are also a good option. Both are best suited to growing
plants during their vegetative phase, so you might consider supplementing with
some red light during the flowering phase with warm-spectrum T5s or red LEDs.
However, don't expect the same types of yields from CFL/T5 lamps that you would
get from HID lamps.

LED rigs are good for both growth phases of cannabis. You will need to place
HID LED fixtures a bit further away from your plants than you would CFLs or T5
fluorescents. Also, high-wattage LED rigs will need ventilation or cooling to control
heat, just like HID lamps.

PROPAGATION Seeds, vegetative cuttings, and retail-ready cannabis transplants
available for purchase. Cannabis comes in both male and female plants, each with

its own purpose. Both flower, but only the flower buds of the female are harvested and used commercially, even though the male plants also contain THC and other compounds. If you're harvesting flower buds, you need to grow only female plants, which have fuzzy white or orange pistils emerging from the flowers.

Unfertilized female plants are much more potent than fertilized plants. If you're trying to hybridize and generate new F1 seed, you need male plants to fertilize the female flowers. About 50 percent of your new seed will be male. You can purchase feminized seed to ensure more female plants.

Vegetative cuttings are typically taken from vigorous, healthy F1 hybrid mother plants maintained in a vegetative state (18 to 24 hours of light daily) that are at least 2 months old. You can take cuttings from a flowering female plant, but it may be as long as a month before the cutting reverts to a vegetative phase after it roots. Ultimately, as with any type of plant, your clones are only going to be as good as the mother plants from which they were taken. It can be more challenging to take successful cuttings from auto-flowering types, as they have a shorter life cycle, but it is possible. If you're going to do so, take cuttings only from the lowest branches.

TEMPERATURE Generally cannabis plants prefer temperatures in the 70°F to 85°F (21–29°C) range during the day, when lights are on. Young vegetative plants do best at these warm temperatures. Once plants begin to flower, cool temperatures to the 65°F to 80°F (18–26°C) range. During both the vegetative and flowering phases, ensure at least a 10-degree Fahrenheit temperature drop at night when the lights are off. When temperatures climb higher than 80°F (26°C), plant growth will slow and harvest quality may diminish.

If you provide intense light, however, typically with HID lighting in the 600-watt range or higher, cannabis plants can still thrive at temperatures upward of 85°F (29°C). Just be sure to limit humidity and keep the air moving. If you're using HID lamps in a sealed setting, you will likely need to use air-cooled lamps or provide cooling ventilation to control temperature and humidity. If you're an advanced grower, you can grow cannabis at hot temperatures around 95°F (35°C) if you also provide more intense light levels and low humidity, and supplement with CO_2.

SEASON Warm-season. Grown indoors year-round.

SPACE Different cannabis species and strains will vary in natural height. *Cannabis indica* strains tend to be more compact than *C. sativa* strains. You can train plants to stay under 2 feet tall at blooming, or plants grown outdoors can stretch to taller than your house. Container size will depend on how large you intend to let plants grow. Plan on 2 gallons of pot volume per 12 inches of plant growth.

GROWING MEDIUM Well-draining, rich with organic matter, with a pH range of 6.0 to 7.0. You can also use soilless media such as coir, but you'll need to keep a careful watch on nutrients and pH, which should be in the 5.5 to 6.5 range. Suitable for hydroponic and aeroponic systems, in which you should maintain a pH range of 5.5 to 6.5.

WATER Cannabis needs consistent moisture, but you should allow the top inch or so of the soil to dry between watering. As plants grow, you'll need to water more often. Overwatering can result in wilt and fungal diseases.

FERTILIZATION You will fertilize cannabis differently depending on whether it's in a vegetative or flowering phase. Vegetative plants need high levels of nitrogen, medium levels of phosphorus, and high levels of potassium. Once flowering, plants need low levels of nitrogen, medium to high levels of phosphorus, and high levels of potassium. If your potting mix is rich in organic matter, you may not need to fertilize in the first few weeks. If you are using a soilless mix or hydroponics, you will need to start fertilizing as soon as seedlings put on their first true leaves. Avoid slow-release fertilizers, as you may get too much nitrogen in the flowering phase. Many fertilizers are formulated specifically for cannabis production.

PESTS & DISEASES Cannabis can fall victim to many pest and disease problems. The most common for indoor plants are aphids, fungus gnats, thrips, spider mites, whiteflies, powdery mildew, root rot, and stem rot.

Carrot

Daucus carota subsp. *sativus*

Carrots are biennial herbs with edible roots that form in their first season of growth. Plants produce a profusion of top leafy growth as well as large umbel flowers that are attractive to pollinators.

PHOTOPERIOD Facultative long-day. While daylengths longer than 12 hours can encourage flowering, warm temperatures may do the same. Grow carrots with 10 to 12 hours of light at cool temperatures to discourage early bolting.

LIGHT REQUIREMENTS Bright light. Outdoors carrots need full sun for 6 to 8 hours, and they require intense light indoors. Be careful with lights that generate too much heat. Choose HO T5, CFL, LED, or MH/CMH lamps with climate control.

PROPAGATION Direct seed only into garden or growing container. Sow seeds ⅛ to ¼ inch deep. Once seedlings have emerged, thin them out so they are 1 to 2 inches apart. If you plan to harvest baby carrots, you can keep them closer together.

CARROT

TEMPERATURE Carrots perform best with daytime temperatures between 60°F and 75°F (15–24°C) and night temperatures between 45°F and 50°F (7–10°C). If your home is too warm or you do not air-cool a grow tent, a cool garage with grow lights is a useful location.

SEASON Cool-season. Seed and grow indoors year-round if you can maintain temperatures. Outdoors direct seed carrots in early spring when soil temperatures are between 40°F and 50°F (4–10°C) or in fall in warm climates.

SPACE Plant in containers at least 12 inches deep. Each carrot will need about 4 square inches of space.

GROWING MEDIUM Can tolerate a variety of soils and pH levels in the 6.5 to 7.8 range, but should be fine, loose, well-draining. Chunky or rocky soil will result in deformed roots.

WATER Keep consistently moist during germination and growth, but take care not to overwater carrots in containers. Pots should never be soggy. Allow containers to drain thoroughly or carrot roots will rot.

FERTILIZATION Apply a balanced granular fertilizer to soil prior to planting. Apply two side-dressings of the same fertilizer during growing phase.

PESTS & DISEASES Powdery mildew, leaf blight, fungus gnats.

VARIETIES & CULTIVARS 'Thumbelina', 'Little Finger', 'Scarlet Nantes', 'Chantenay', 'Danvers Half Long'

HARVEST You can harvest baby carrots when they are small or allow roots to develop to full size. Don't leave carrots in pots (or in the ground) for too long or they will begin to crack and become tough.

GROW THE SAME WAY Parsnips

Chives

Allium schoenoprasum

Chives are resilient, easy-to-grow herbs. They are small bulbs that belong to the onion family and serve as ornamental blooming perennials. Chives are grown primarily for their garlic- or onion-flavored foliage.

PHOTOPERIOD Obligate short-day. Plants can go dormant with daylengths of less than 14 hours. Grow plants indoors with 16 to 18 hours of light to prevent bolting.

LIGHT REQUIREMENTS Bright light indoors. In an outdoor garden, plants need 6 to 8 hours of direct full-sun exposure.

In hot climates, plants can tolerate some afternoon shade. Choose cool-spectrum HO T5, CFL, LED, and MH/CMH lamps with climate control.

PROPAGATION Direct seed, transplants, divisions. Sow seeds on surface of soil or barely cover. Seeds germinate well at temperatures between 60°F and 85°F (15–29°C).

TEMPERATURE Chives grow well between 50°F and 75°F (10–24°C), and will flower more in hotter temperatures.

SEASON Set outdoors any time during the growing season or grow indoors year-round.

SPACE Individual plants are small. Grow in 1- to 5-gallon containers.

GROWING MEDIUM Tolerates a wide variety in the garden and in containers; well-draining in the 6.0 to 7.0 pH range.

WATER Moderate to low needs and moderate to low humidity requirements. Plants do not tolerate soggy or waterlogged soils.

FERTILIZATION Chives are moderate to high nitrogen feeders. If you add composted manure or a high-nitrogen fertilizer to the soil at planting time, you likely don't have to provide supplemental fertilization. If you harvest from your plants frequently, side-dress them with additional composted manure or fertilizer once or twice a year.

PESTS & DISEASES Downy mildew under cool, humid conditions.

VARIETIES & CULTIVARS Common chives, garlic chives, blue chives (Siberian), giant Siberian chives

HARVEST Snip any size leaves at the base. Store in a glass of water on the counter or in the refrigerator.

TIPS After three years, you may need to divide your clump or container of chives. In the garden, plants can spread aggressively by seed.

Cilantro, Chinese parsley

Coriandrum sativum

The pungent leaves of cilantro brighten the flavors of many cuisines. The seeds of cilantro plants are called coriander. Left to flower in the garden, plants are attractive ornamentals and good food for local bees.

PHOTOPERIOD Obligate long-day. Grow with 10 to 11 hours of light to keep plants vegetative for as long as possible. If you want plants to flower for harvesting coriander

CILANTRO, CHINESE PARSLEY

seed, extend the daylength and raise the temperature. Once plants are triggered to flower, the plant will begin to die after it sets seed.

LIGHT REQUIREMENTS Moderate for harvesting foliage, bright for flowering. You can grow in partial shade outdoors. Choose cool-spectrum HO T5, CFL, LED for leaves; warm-spectrum fluorescent, LED, MH/CMH, or HPS with climate control to transition plants to flowering and seed production.

PROPAGATION Direct seed into final container and garden. Sow seed ¼ to ½ inch deep at soil temperatures between 55°F and 65°F (13–18°C). Transplants during the warm season will often bolt quickly.

TEMPERATURE Cilantro is particularly heat-sensitive and does best at growing temperatures of 50°F to 75°F (10–24°C). Keep on the cooler side to encourage more foliage. Plants scramble to bolt the moment temperatures rise above 70°F (21°C) with lengthening days. Keep cilantro in a cool house or garage, or in an air-cooled grow tent.

SEASON Cool-season. In warm climates, cilantro can be grown outdoors only in the cooler fall and winter months. In cool climates, it can be grown in spring and early summer. Grow year-round indoors under cool conditions.

SPACE Plants are compact when vegetative, but can reach several feet tall when in flower. Space plants 6 to 8 inches apart. Grow in 6-inch diameter or 2-gallon containers.

GROWING MEDIUM Well-draining, with a pH range of 6.0 to 7.0. Plants are also suited to hydroponic or aeroponic systems.

WATER Moderate to light needs, but keep soil evenly moist. Plant can be triggered to flower if you allow soil to dry repeatedly under long days.

FERTILIZATION Cilantro is not a heavy feeder and won't require much fertilization. Add some compost or composted manure to your growing media before seeding. If you harvest regularly, feed plants with a half-strength solution of liquid humus, seaweed, or fish emulsion twice per month.

PESTS & DISEASES Aphids, whiteflies, and fungus gnats will be most common indoors.

VARIETIES & CULTIVARS 'Slow Bolt', which is less sensitive to photoperiod and temperature, 'Long Standing', 'Leisure'

HARVEST You can snip stems and leaves as you need them, or harvest entire plants if you have enough succession plantings. It's best to use fresh cilantro, as dried loses most of its flavor. Store in a glass of water on the counter or in refrigerator. To harvest coriander seeds, allow plants to flower and seeds to dry on the plant.

GROW THE SAME WAY Flat leaf and curled leaf parsley, anise, culantro (*Eryngium foetidum*), chervil, celery, cutting celery

Citrus

Lemon

Citrus limon

VARIETIES & CULTIVARS 'Dwarf Meyer', 'Dwarf Meyer Improved', 'Eureka', 'Variegated Pink', 'Ponderosa'

Lime

Citrus spp.

VARIETIES & CULTIVARS Indonesian or Thai (*Citrus hystrix*), Dwarf Key (Mexican) and Thornless Mexican lime (*C. aurantifolia*), Persian lime (*C. latifolia*)

Orange

Citrus ×sinensis

VARIETIES & CULTIVARS 'Cara', 'Valencia', 'Trovita'

Mandarin

Citrus reticulata

VARIETIES & CULTIVARS 'Owari', 'Gold Nugget'

TOP 'Meyer' lemon in flower.

BOTTOM 'Ponderosa' lemon.

There are many different species and cultivars of citrus, all with wonderfully fragrant flowers. If you're lucky enough to live in a warm, frost-free climate, you can grow citrus outdoors. But in cold climates or even mild areas that experience a handful of winter frosts, grow citrus trees in containers and cover or bring indoors during cold snaps.

While citrus is easy to grow in containers, they hate living indoors. When you see photos of a beautiful fruit-filled lemon tree next to a small indoor window, know that it most likely did not grow and fruit in that location. If you happen to have a very large south- or southwest-facing window that gets direct full sunlight for more than 6 hours a day, you might be able to keep your citrus happy. But most indoor gardeners will require bright supplemental light.

PHOTOPERIOD Day-neutral. Cooling temperatures and less water will trigger flowering once plant stems are mature enough to flower.
LIGHT REQUIREMENTS Bright light indoors, with direct full-sun exposure for 8 to 12 hours outdoors. Grow citrus indoors with 12 to 16 hours of supplemental

light. You can get good growth and vigor from citrus using MH/CMH, CFL, or intense HO T5 lamps, but you'll need a full-size grow lamp. One small CFL spotlight will not support your citrus trees. HID lamps are ideal for citrus, but most indoor gardeners do not want a big HID lamp hanging in the living room. I bring my four dwarf citrus trees indoors for the winter and maintain them under a large eight-lamp HO T5 fluorescent fixture in my garage. You can also place citrus plants inside grow tents with HID lamps for best production. If you have very bright south-facing windows, you may be able to combine that natural light with a few spotlight CFLs to keep a group of citrus happy in your living room.

PROPAGATION Buy nursery-grown plants or propagate vegetative cuttings. You can start citrus from seeds you kept from fruit you purchased. If you start a seed from a hybrid variety of citrus, your seedlings won't come true to variety. Plus, it takes years to grow a mature citrus from seed. If you're really patient, go for it. You can also propagate citrus from cuttings, air-layering, and grafting. Otherwise, pick up young or mature specimens at your local nursery and pot them. Note that there are shipping restrictions for citrus plants to certain states to prevent the spread of pests and diseases to agricultural crops.

TEMPERATURE The ideal temperature range for most citrus is between 70°F and 90°F (21–32°C). A temperature drop during the winter months typically triggers blooming. Warm temperatures in summer help fruit develop a sweet flavor. Most citrus can be damaged at temperatures below 27°F (–3°C),

TOP LEFT Key limes.

TOP RIGHT 'Meyer' lemons.

BOTTOM LEFT Oranges.

although some citrus types can survive lower temperatures. Rapid swings within the ideal temperature range can cause issues such as fruit drop or fruit splitting.

SEASON Year-round outdoors in frost-free climates. Indoors year-round if provided with enough supplemental light. A good compromise is to grow containerized citrus outdoors during the warm seasons, then maintain them indoors through the winter months.

SPACE Standard citrus trees can grow 15 to 20 feet tall. For indoor gardening, it's best to stick with dwarf types that you can maintain at 3 to 6 feet tall. Young nursery plants can go into 12-inch diameter or 5-gallon containers. As plants get larger, you can bump them up to a larger container with a diameter of 14 to 20 inches.

GROWING MEDIUM In landscape, citrus prefer well-draining, sandy loam type with a pH range between 5.5 and 6.5. For container growing, choose organic outdoor-weight, then add coir to make up about one-third of the soil mix.

WATER Water citrus consistently, but do not let soil get waterlogged. Make sure pots can drain thoroughly, and never allow citrus to sit in trays full of water. Yellowing leaves can indicate poor drainage or a nutrient deficiency.

FERTILIZATION Citrus trees are heavy nitrogen feeders. Choose a fertilizer with a 2:1:1 or 3:1:1 ratio. There are special fertilizer blends labeled for citrus plants. Apply a fertilizer to the soil at planting time, then fertilize monthly during the active growing season (late winter through summer) and once every two months during fall and early winter. Foliar applications of liquid kelp or seaweed can also be helpful.

PESTS & DISEASES While citrus can experience many issues when grown outdoors, you'll find scale, whiteflies, and aphids to be your biggest pests indoors.

GROW THE SAME WAY Blood orange, grapefruit, kumquat, finger limes, calamondin (×*Citrofortunella microcarpa*)

TIPS Hand-pollinate flowers if your citrus plants are blooming indoors.

TOP Satsuma orange.

BOTTOM Citrus flowers are both beautiful and incredibly fragrant.

Cucumber

Cucumis sativus

Believed to be native to India, cucumbers have been cultivated for thousands of years. They are a versatile, healthy, delicious fruit.

PHOTOPERIOD Day-neutral. Cucumbers will flower with a variety of photoperiods and you can grow them successfully with anywhere from 12 to 18 hours of light. I find 12 to 14 hours to be adequate with high-intensity lighting.

LIGHT REQUIREMENTS Bright light indoors, with direct full-sun exposure for 6 to 8 hours outdoors. HO T5, LEDs, MH/CMH, or HPS combined with blue light from HO T5 or LEDs. DLI 20 to 30 mol m^{-2} d^{-1}.

PROPAGATION Direct seed into the garden and the final growing container indoors. Cucumber seedlings do not transplant well. Seeds can germinate at temperatures between 60°F and 90°F (15–32°C), but the optimum range is 80°F to 90°F (26–32°C), as seeds can rot at cooler temperatures. Place a seed heat mat under containers when germinating.

TEMPERATURE Cucumbers can handle warm conditions, but perform best at temperatures between 70°F and 75°F (21–24°C), so normal household temperatures are ideal. If you're growing cucumbers inside a grow tent and temperatures are warmer during the day, be sure night temperatures are cooler.

SEASON Warm-season. Direct seed outdoors after all chance of frost has passed and soil temperatures have warmed. Grow indoors year-round.

SPACE Standard cucumbers produce vines that can reach 20 feet tall, but bush cucumbers will stay in the 3- to 5-foot range and are easy to keep in containers indoors. Choose a container at least 8 inches wide and about 12 inches deep.

GROWING MEDIUM Loose, well-draining, fertile with a pH range of 5.5 to 7.0. Plants are also suited to hydroponic, aquaponic, and aeroponic systems.

'Martini' cucumber.

WATER Cucumbers produce fruit that contain a lot of water, so they must stay consistently moist for fruit to develop properly. If you let your cucumbers go dry, you may see flower drop or your fruit might have a tough texture. Inconsistent watering can lead to blossom end rot, a deformity caused by a disruption in calcium uptake. Plan to water your cucumbers daily once they reach flowering and fruiting size.

FERTILIZATION Cucumbers are heavy feeders. Add a balanced granular vegetable fertilizer to the container at seeding time. As vines begin to grow vigorously, you can feed weekly with a balanced organic liquid fertilizer, or side-dress with a granular several times until harvest. You can also apply a foliar spray of fish emulsion several weeks after seedlings emerge.

PESTS & DISFASES Cucumber bcctles and squash vine borers are dreaded pests in the outdoor garden, but you probably won't have to deal with them indoors. Powdery mildew, downy mildew, and other fungal leaf spot diseases are common; whiteflies and spider mites on stressed plants can also be an issue.

VARIETIES & CULTIVARS 'Bush Pickle', 'Spacemaster', 'Salad Bush'. Burpless cucumbers (which contain little to no cucurbitacin) have thinner skins and sweeter flavors, and they can be easier to digest than other cucumbers.

HARVEST Harvest cucumbers when they have reached the mature size described on the seed packet. Don't leave the fruit on the vine too long or the seeds inside will grow large and inedible.

GROW THE SAME WAY Mexican sour gherkin, dwarf melons

TIP Indoor gardeners will have to hand-pollinate. Cucumbers and other cucurbits produce male and female flowers on the same plant. Male flowers typically emerge first, followed by female flowers, which have a swollen ovary (looks like a mini fruit) behind the flower.

Flowers on a cucumber plant.

Dill

Anethum graveolens

Use fresh dill on vegetables and dishes, or add to pickles and preserves. When left to flower in the garden, plants make beautiful edible ornamentals that attract many beneficial insects.

PHOTOPERIOD Obligate long-day. Grow with 10 to 11 hours of light to suppress flowering and extend foliage production.

LIGHT REQUIREMENTS Medium. Outdoors plants can produce well in partial shade. Cool-spectrum HO T5, CFL, LED are best. MH/CMH with climate control. DLI 12 to 16 mol m^{-2} d^{-1}.

PROPAGATION Direct seed into final growing container or garden. Sow seeds on the soil surface or just barely cover with soil at temperatures between 60°F and 70°F (15–21°C).

TEMPERATURE Dill grows best at cool temperatures between 60°F and 80°F (15–26°C).

SEASON Cool-season. Plant in late summer, fall, or late winter. Grow indoors year-round with climate control.

SPACE Dill plants need a bit of room, with dwarf varieties reaching 18 inches tall and standard varieties stretching to 3 to 4 feet tall. Use 1- to 3-gallon containers for mature plants.

GROWING MEDIUM Well-draining with a pH range of 5.5 to 7.5. Plants are also suited to hydroponic and aeroponic systems.

WATER While dill can tolerate some dry conditions, you'll get much better production if you water regularly and consistently. Moderate humidity.

FERTILIZATION Dill does best with higher levels of N-P-K. Add compost or composted manure to potting media at planting, then liquid feed once a month with a half-strength solution of liquid humus, seaweed, or fish emulsion once per month.

PESTS & DISEASES Pests are not common on dill, especially indoors, other than occasional aphids. Outdoors, several types of caterpillars feed on dill foliage.

VARIETIES & CULTIVARS 'Dukat', 'Fernleaf', 'Hercules'

HARVEST Start snipping leaves as soon as they unfurl. Harvest stems from outside the central growing point, which allows new leaves to emerge from the center. Store in a glass of water on the counter or in the refrigerator.

GROW THE SAME WAY Green fennel, bronze fennel

TIPS Dill will sprint to flower at warmer temperatures and long days. You can seed succession crops every few weeks to keep a steady supply of foliage.

Kale

Brassica oleracea

Kale, or leaf cabbage, refers to several varieties within the species *Brassica oleracea*. Many are beautiful ornamentals that can serve dual purpose in an edible landscape, offering a variety of foliage colors and textures.

PHOTOPERIOD Day-neutral. You can grow kale and related leafy crops with longer photoperiods of 14 to 16 hours to boost production.

LIGHT REQUIREMENTS Medium. In the outdoor garden, kale and related crops can tolerate some shade. Indoors, cool-spectrum HO T5, CFL, and LEDs are sufficient.

PROPAGATION Grow transplants from seed. Seed can germinate at temperatures between 40°F and 85°F (4–29°C), with optimal germination at about 70°F (21°C).

TEMPERATURE Kale prefers cool conditions. Plants will bolt once temperatures reach about 80°F (26°C). Grow kale indoors in a space that stays consistently cool, such as a basement or a garage. You will need to cool air inside a sealed grow tent.

SEASON Kale are typically cool-season crops planted outdoors in the fall, and again through winter in warmer climates. Grow indoors when you can control cool temperatures.

SPACE Kale plants typically grow from about 1½ to 3 feet tall and wide, depending on the variety. However, their root systems are generally compact, so you can grow them in pots ranging from 6 to 8 inches wide and 8 to 10 inches deep. Use larger containers to grow multiple plants together.

GROWING MEDIUM Fertile, with organic matter and a pH range of 6.0 to 7.5. Plants are suited to hydroponic, aquaponic, and aeroponic systems.

WATER While kale is hardy and can tolerate some drought, watering plants consistently will produce the best yields. Drying out can result in tough leaves with less flavor. Do not overwater or let plants sit in water.

FERTILIZATION Kale is not a heavy feeder, so if your soil is rich in organic matter, you may not have to provide supplemental fertilization. It is a good idea to apply a granular balanced vegetable fertilizer to the container at the time of transplanting.

PESTS & DISEASES While cabbage loopers, cut worms, flea beetles, slugs, and a few other critters can be a problem for kale in the outdoor garden, you probably won't experience these pests indoors (unless you buy transplants that carry eggs). Soil-borne fungal diseases are possible, but good water management indoors will minimize these issues. Aphids, fungus gnats, and spider mites will more likely be an issue indoors.

VARIETIES & CULTIVARS 'Dinosaur', 'Winterbor', 'Red Russian', 'Vates Blue Curled'

HARVEST You can harvest kale leaves continuously by cutting leaves at the base of the stem. Harvest older leaves first, which allows the new growth to emerge from the center of the plant.

GROW THE SAME WAY Brussels sprouts, cabbage, chard, collard greens; chard is the most heat-tolerant, and you can grow it through the summer months outdoors or in warmer indoor growing environments.

Lettuce

Lactuca sativa

Lettuce plants are annual cool-season leafy vegetables categorized in four principal types butterhead (bibb), loose-leaf, crisphead (iceberg), and romaine (Cos). Crisphead is the most difficult for home gardeners to grow because of its sensitivity to temperature.

PHOTOPERIOD Facultative long-day. Grow lettuce plants with 12 hours of light to suppress bolting and extend your loose-leaf harvesting. If your plants are still bolting with 12 hours of light, temperatures are likely too warm.

LIGHT REQUIREMENTS Medium. Outdoors plants can produce well in partial shade. Cool-spectrum HO T5, CFL, LED are best. MH/CMH with climate control. DLI: 12 to 16 mol m^{-2} d^{-1}.

PROPAGATION Direct seed or transplants. Grow new seedlings with 14 to 16 hours of light placed 2 to 3 inches from the seedlings. Heat mats are not necessary. As seedlings put on two to three true leaves, reduce light photoperiod to 12 hours.

TEMPERATURE Cool-season. Seeds germinate best at 70°F (21°C), but can germinate at temperatures between 50°F and 80°F (10–26°C). Growing plants perform best at temperatures of 60°F to 70°F (15–21°C), but if your indoor temperatures are a bit higher, just make sure night temperatures drop about 10 degrees Fahrenheit.

SEASON Fall through spring outdoors, year-round indoors if you can provide a good temperature range.

SPACE Lettuce is perfect for small spaces. Shallow 4- to 6-inch containers with drainage.

SOIL Use a lightweight seed-mix potting soil, especially if you direct seed into the final container. If you use a heavier, organic, compost-based mix, lighten it up with some coir. Plants are also well suited to hydroponic, aeroponic, and aquaponic systems.

WATER Lettuce plants comprise mostly water and have shallow root systems, so keep soil evenly moist. Soggy soil will cause rot.

FERTILIZATION Apply a balanced vegetable fertilizer to the soil at planting. Fertilize after the first set of true leaves emerge, using a half-strength solution of liquid humus, seaweed, or fish emulsion, and repeat weekly.

PESTS & DISEASES Damping off, powdery mildew, fungus gnats, aphids, whiteflies.

VARIETIES & CULTIVARS 'Lollo Rossa', 'Oak Leaf', 'Black Seeded Simpson', 'Buttercrunch', 'Red Sails', 'Adriana', 'Jericho', 'Magenta', mesclun loose-leaf mixes

HARVEST For continual leaf harvesting, snip leaves from outside the central growing point, which allows new leaves to emerge from the center. If you plan to harvest an entire head, you can grow under longer photoperiods, but be sure to harvest the plant as soon as it fills out, before it bolts.

GROW THE SAME WAY Chinese cabbage, Napa cabbage, endive, escarole, frisée, radicchio, red orach, cress, watercress, mustard greens, Asian greens

TIPS Lettuce foliage becomes bitter and plants stop producing new leaves once they bolt. You can mix together any type of lettuce seed and other greens to create your own mixes or for use as microgreens.

Microgreens
Many assorted species

Microgreens are the seedling sprouts of many different types of vegetables and herbs, including spinach, kale, chard, beets, arugula, basil, lettuce, broccoli, cilantro, and others. The difference between microgreens and sprouts is that microgreens are a bit older than sprouts and have developed at least a couple of true leaves and a small root system.

PHOTOPERIOD Varies depending on species, but most will be long-day. You will harvest microgreens when they are very young, however, before any flower initiation should begin. If you snip them in a timely fashion, you can grow them with longer photoperiods of 14 to 16 hours, depending on your lamp's light output, to boost production.

LIGHT REQUIREMENTS Medium to bright. Grow microgreen seedlings close to the light source, about 4 inches away. HO T5, CFL, and LED lamps are best, as HID lamps will produce too much heat for the small seedlings. DLI: 6 to 18 mol m^{-2} d^{-1}. You can grow microgreens successfully using a wide DLI range. Microgreens grown at higher DLIs contain more beneficial anthocyanin pigments, but you will use more electricity to produce them.

PROPAGATION Direct seed into enclosed seed tray without drainage holes or any shallow container without a hole. Sprinkle seeds across surface of moist soil and mist to moisten. Create a thick stand of seedlings, but don't seed so densely that seeds are on top of one another; this will cause air circulation problems.

TEMPERATURE The ideal temperature for most microgreens is 67°F to 70°F (19–21°C). If your room is cold, use a heat mat under the seed tray to speed up germination.

SEASON Year-round indoors. You can also grow microgreens outdoors during each crop's growing season.

SPACE Minimal. You need only about 2 square feet of space to grow a full tray of microgreens. Great crop for countertops.

GROWING MEDIUM Lightweight seed-starting mix; 2 inches of soil within the tray is plenty. You can also use felt-like water-wicking mats in place of soil. Simply place the mat in the bottom of your tray and sprinkle seeds on top.

WATER Moisten soil or seed mat before seeding. Use a humidity tray until seeds sprout, then remove it. Keep soil consistently moist (but not soggy) with a spray bottle.

FERTILIZATION Mist microgreens with a half-strength organic liquid fertilizer, such as liquid seaweed or kelp, just as they begin to develop a true leaf. This will help improve flavor and nutritional value.

PESTS & DISEASES Mold is the most common problem if seedlings stay too wet or humidity is too high. Seeding too densely can encourage a mold problem, rendering your microgreens inedible. Fungus gnats can be an issue in soil-based media.

VARIETIES & CULTIVARS Any type of salad or leafy green or vegetable plants with edible foliage.

HARVEST Cut the seedlings using sharp snips at the base of the stem, above the roots. Harvest after the seedlings have developed only a few true leaves, which typically occurs in 9 to 14 days.

TIP Do not use tomatoes, peppers, or eggplant seeds for microgreens, as these nightshade family members produce toxins in their leaves and stems.

Peppermint, Spearmint

Mentha piperita, *M. spicata*

There is nothing more refreshing than fresh mint. Its aromatic and flavorful foliage enhances teas, beverages, desserts, and many mealtime dishes. Mint is a shade-tolerant perennial herb that is easy to grow indoors.

PHOTOPERIOD Long-day. Mint grows more vigorously and produces more foliage and essential oils with longer days. You can grow with 12 to 13 hours of light to slow flowering, or keep flowers pinched and harvest regularly under longer photoperiods.

LIGHT REQUIREMENTS Medium to low. Outdoors plants can produce well in partial shade. Cool-spectrum HO T5, CFL, LED are best. MH/CMH with climate control. You can grow mint in containers indoors in a bright sunny window, but it will get leggy if the room is not very bright. DLI: 10 to 16 mol m^{-2} d^{-1}.

PROPAGATION Direct seed or transplants. Seed germinates best at temperatures between 68°F and 75°F (20–24°C). Mint also grows easily from cuttings.

TEMPERATURE Mint is at its best at temperatures between 65°F and 70°F (18–21°C), and is usually happy in typical indoor home temperatures. Mint can tolerate much warmer temperatures with less intense direct light.

SEASON In many climates, mint is a hardy perennial spreader outdoors and can be planted out any time during your normal growing season. Grow indoors year-round.

SPACE Mint is an aggressive grower that will creep and spread through underground runners with a shallow but wide root system. Plants can grow 12 to 24 inches tall, depending on variety. You can use many different-size containers to grow mint, but wider is better.

WATER Mint likes moisture and will quickly wilt when the soil begins to dry. The smaller your container, the more often you will have to water mint to keep it healthy.

FERTILIZATION Mint is an aggressive grower without much fertilization. In containers you can feed plants with a half-strength solution of liquid humus, seaweed, or fish emulsion once per month.

PESTS & DISEASES Outdoors, mint can succumb to soil-borne fungal diseases such as verticillium wilt and anthracnose, as well as to spider mites and root borers. These will be less of an issue indoors, but whiteflies and spider mites can be problem. Occasionally your plants may have aphids.

VARIETIES & CULTIVARS There are hundreds of mint varieties, but popular types include chocolate mint, orange mint, lavender mint, apple mint, and pineapple mint.

HARVEST Snip shoots of mint as you need them, fresh or dried. Store in a glass of water on the counter or in the refrigerator.

GROW THE SAME WAY Lemon balm, catnip, catmint

TIPS If plants get leggy, cut back the entire plant to just above the growing crown to refresh and encourage healthier new growth.

Oregano, Wild Marjoram

Origanum vulgare subsp. *hirtum,*
O. vulgaris, O. rotundifolium

A hardy and easy-to-grow perennial herb, oregano provides the signature flavor for many Italian and Mediterranean dishes. Very easy to grow in containers for continuous harvest. Some oregano types produce large, showy flowers, making them excellent edible ornamentals for the garden.

PHOTOPERIOD Obligate long-day. Grow with 12 hours of light to suppress flowering and encourage leafy growth.

LIGHT REQUIREMENTS Bright light indoors. Plants do best with 6 to 8 hours of full direct sun outdoors. Cool-spectrum HO T5 fluorescent, CFL, or LED. Use MH/CMH lamps for more intense production.

PROPAGATION Direct seed, transplants, and cuttings. Sow seed on surface of soil or barely below the surface at temperatures of 60°F to 70°F (15–21°C). Transplant outdoors or to final containers indoors. You can take cuttings from established plants and easily root them.

TEMPERATURE You can grow oregano at temperatures between 55°F and 80°F (13–26°C), but it can also tolerate warmer temperatures and often thrives in hot climates and hot grow tents.

SEASON You can plant perennial oregano any time during your outdoor growing season, and it will remain evergreen in warm climates. Grow indoors year-round.

SPACE Oregano typically grows up to 2 feet tall and 2 to 3 feet wide. It will often trail down the sides of the container. Easy to grow in containers, but plants have a good-size root system. Use 2- to 5-gallon containers.

SOIL Oregano needs very good drainage. Loose, well-draining, with a pH range of 6.0 to 7.5. You can loosen the mix with coir or perlite. Plants are suited to hydroponic, aquaponic, and aeroponic systems.

WATER Oregano is drought tolerant, so do not overwater. Grow in low humidity, and allow plants to drain thoroughly and dry a bit between watering. Wet conditions can cause fungal disease problems.

FERTILIZATION Apply a granular balanced herb or vegetable fertilizer at planting time. For ongoing care, feed plants with a half-strength solution of liquid humus, seaweed, or fish emulsion twice per month.

PESTS & DISEASES Powdery mildew, root rot, rust, aphids, spider mites.
VARIETIES & CULTIVARS Greek and Italian for strongest flavors. Ornamentals with big blooms, such as 'Kent Beauty' and 'Amethyst Falls', will have very mild flavors.
HARVEST Harvest oregano regularly by snipping off tip sections of stems. Use fresh or dry for storage. Oregano is a vigorous grower and will benefit from a severe cutting back if plants get leggy.
GROW THE SAME WAY Rosemary, sage, marjoram, bay

Peas

Pisum sativum, snow pea (*P. sativum* var. *saccharatum*), sugar snap pea (*P. sativum* var. *macrocarpon*)

Peas are one of the oldest cultivated vegetables (although botanically the pods are a fruit and the peas are seeds). Peas are a very versatile crop: you can grow peas for the fresh green seeds, the green pods, foliage, young shoots for sprouts and micro-greens, and dried seeds. All are edible with a fresh, bright flavor.

PHOTOPERIOD Facultative long-day. Grow with 12 hours of light to boost vegetative growth, then switch to 14 to 18 hours of light for flowering and fruiting.
LIGHT REQUIREMENTS Bright to medium, with 6 to 8 hours of sun preferred outdoors. Peas can tolerate some shade, especially in hot climates, but more light delivers better yields. Indoors you want bright light. Intense HO T5, CFLs, or LEDs. MH/CMH with climate control.
PROPAGATION Direct seed outdoors or into final growing container indoors. Pea seedlings do not transplant well. Plant seeds 1 to 2 inches below the soil. Seeds can germinate at temperatures between 40°F and 85°F (4–29°C), but will take longer at the cooler temperatures. However, while warmer soil temperatures can speed up germination, the young seedlings that emerge often don't tolerate the extra warmth. Skip the heat mat when germinating peas.
TEMPERATURE Peas grow best between temperatures of 50°F and 70°F (10–21°C). They can tolerate warmer temperatures, but you will need to water more frequently.

SEASON Cool-season. Young plants will tolerate light frosts. Outdoors, sow seeds 4 to 6 weeks before your last frost date. In hot climates, seeding in late summer or early fall is best for a good spring harvest. Grow indoors year-round if temperatures aren't too warm.

SPACE Peas are vining plants that grow quickly up to 8 feet and ½ to 1 feet wide, and they need a support structure. Many compact dwarf peas grow 1 to 2 feet tall and wide, making them perfect for indoor container culture in small spaces. Use 1- to 3-gallon containers. You can seed several plants in a larger container.

GROWING MEDIUM Peas can tolerate different types of soil, but do best in loose, sandy, well-draining, with a pH range of 6.0 to 7.5. Use a potting mix with plenty of organic matter and add some coir to loosen it. Plants are suited to hydroponic, aquaponic, and aeroponic systems.

WATER Peas grow quickly and need consistent watering throughout the growing cycle. Don't let plants dry too much between watering, but take care not to overwater.

FERTILIZATION Peas do not respond well to overfertilization, and too much nitrogen will result in limited fruit. Add granular vegetable fertilizer with a 1:2:2 ratio at planting time, then reapply once plants begin to set fruit. Peas form beneficial relationships with soil bacteria, *Rhizobium* spp., to take up nitrogen from the soil. To add these bacteria to soil, look for packets labeled *Garden Inoculant*. You can also pretreat your seeds with inoculant.

PESTS & DISEASES Many pest and disease problems can affect peas in the out-door garden, but indoor issues will be limited to a few specific culprits, such as damping off, crown rot, aphids, thrips, powdery mildew, and downy mildew.

VARIETIES & CULTIVARS 'Patio Pride', 'Alaska', 'Little Sweetie', 'Sugar Ann', 'Sugar Snap', 'Snowbird', 'Dwarf Grey Sugar', 'Oregon Sugar Pod II', 'Dual', 'Garden Sweet'

HARVEST English peas are typically shelled. Harvest pods when you see the rounded peas swelling but the pod is still tender, then remove peas. Snow peas are harvested when the pods are young and flat. You can harvest sugar snap types when they are young and flat, like snow peas, or leave them to mature longer to be shelled like English peas.

Peppers

Capsicum annuum, C. chinense, C. baccatum,
C. frutescens, and *C. pubescens*

Peppers, nightshade family relatives of tomatoes and eggplant, are a generally compact and easy-to-grow crop for the outdoor and indoor garden. Hot peppers are some of the easiest edibles to grow, and I highly recommend them as a first crop for new gardeners. Endless varieties of peppers offer a multitude of colors, sizes, and flavors.

PHOTOPERIOD Day-neutral. You can grow under a wide range of photoperiods, but unlike tomatoes, you can boost pepper production with a longer daylength of 20 hours without damaging plants. If you're using less intense lights, grow with 14 to 20 hours of light. If you're using HID lamps that deliver more light, you can grow with shorter photoperiods of 12 to 14 hours.

LIGHT REQUIREMENTS Bright light indoors, with direct full-sun exposure for 6 to 8 hours outdoors. Intense HO T5, CFL, LED, MH/CMH; or HPS combined with blue light from HO T5 or LEDs, spectrum-enhanced HPS, or Dual Arc. DLI: 20 to 30 mol m^{-2} d^{-1}.

PROPAGATION Sow seeds indoors and transplant outdoors after the last frost when the soil is warm, or sow seeds indoors and transplant into final growing container indoors. Sow seeds ¼ inch below the soil. Use a heat mat to keep soil temperature around 80°F (26°C). Pepper seed will not germinate if soil temperatures drop to 55°F (13°C). You can also propagate peppers by cuttings and grafting.

TEMPERATURE Peppers need consistently warm temperatures between 65°F and 80°F (18–26°C). However, hot peppers can tolerate temperatures of 90°F (32°C) and hotter and still produce plenty of fruit. This makes them easy to grow in a very warm grow tent. Bell peppers, however, prefer more moderate, consistent temperatures.

SEASON Warm-season. Place seeds and transplants outdoors once day and night temperatures are warm enough in late spring or summer. Grow indoors year-round.

SPACE Small to large, depending on variety. Compact varieties can be grown in 6-inch diameter or 1-gallon containers, while you can transplant larger types into 3- to 5-gallon containers. Peppers will generally grow 1½ to 3 feet tall. Some varieties are upright, while others have a wide, sprawling growth habit.

SOIL Peppers prefer soil that is loose and rich in organic matter. Plants are suited to hydroponic, aquaponic, and aeroponic systems.

WATER Consistent watering is needed, but peppers can tolerate some drying between watering. Peppers do not tolerate overwatering and should never have standing water in their water trays. Peppers are sensitive to cool, moist soil, and young plants can rot in these conditions.

FERTILIZATION Use a balanced vegetable fertilizer at planting time, then wait until plants develop small fruit and fertilize again. You can use liquid humus, kelp, or seaweed as a soil drench or foliar spray if plants appear nutrient-deficient. Too much nitrogen can limit flowering and weaken plants. You can also spray peppers with a calcium phosphate solution as they mature to reduce blossom end rot or flower drop.

PESTS & DISEASES Indoors, aphids are the biggest culprit on young pepper foliage. Whiteflies and thrips can also be an issue, as can blossom end rot. Pests on peppers are less common outdoors, but flea beetles can still be an issue indoors, especially if you buy transplants.

VARIETIES & CULTIVARS Jalapeño, relleno, poblano, serrano, cayenne, sweet bell, pimento, banana, habanero, ghost

HARVEST You can harvest peppers at different stages of ripeness, from green to red to dry on the plant. Most peppers change in color from green to purple and then mature to yellow, orange, or red. Green bell peppers are simply unripe. More mature fruit will have a sweeter flavor. Some varieties remain green at maturity but will be much darker in color.

GROW THE SAME WAY Eggplant, tomatillo

TIPS Peppers (and eggplant) are wind-pollinated. You can boost fruit set by gently shaking the plants to move pollen around. Running fans in your growing area will also help pollination.

Spinach

Spinacia oleracea

A member of the amaranth family, this annual leafy green is grown for its leaves, which you can cook or eat fresh. Spinach is a low-calorie, nutrient-dense food that is easy to grow indoors and in the garden.

PHOTOPERIOD Long-day, with a critical daylength of 13 hours. Spinach can develop flower buds at shorter photoperiods, but bolting and flowering occurs at photoperiods longer than 13 hours. Grow spinach with 12 hours of light to keep plants vegetative and suppress flowering.

LIGHT REQUIREMENTS Medium. Outdoors plants can produce well in partial shade. Cool-spectrum HO T5, CFL, LED are best. MH/CMH with air cooling. DLI: 12 to 16 mol m^{-2} d^{-1}.

PROPAGATION Direct seed outdoors in the garden 4 to 6 weeks before last frost and 6 to 8 weeks before first frost in fall, or grow transplants from seed indoors for setting in the garden or growing indoors. Sow seed ½ to 1 inch below the soil. Cool soil temperatures between 50°F and 60°F (10–15°C) are best for seed germination. Seeds can germinate at warmer soil temperatures, but germination rates will drop. Skip the heat mat when germinating spinach seed.

TEMPERATURE Spinach grows best in cool temperatures between 40°F and 75°F (4–24°C). Warmer temperatures can hasten bolting. If your indoor temperatures are a bit warmer, keep the photoperiod on the short side.

SEASON Cool-season. Seed outdoors through fall, late winter, and early spring, depending on your climate. Grow indoors year-round in cool temperatures.

SPACE Spinach plants are compact, each requiring ½ to 1 square foot of space, and can be grown in 6- to 8-inch diameter or 1- to 2-gallon containers. You can grow multiple plants together in larger containers.

GROWING MEDIUM Sandy, loose, with a pH range of 6.5 to 7.5, but can tolerate a variety of soils if they aren't too acidic. Choose a loose, well-draining mix with organic matter. Plants are suited to hydroponic, aquaponic, or aeroponic systems.

WATER Plants have shallow root systems and need consistent, even watering, but not soggy soil. Plants will wilt quickly when temperature is too warm or if you forget to water.

FERTILIZATION Spinach isn't a heavy feeder and you won't need to fertilize much. Apply a balanced vegetable fertilizer at planting time, then apply liquid humus, kelp, or fish emulsion if plants appear pale.

PESTS & DISEASES Indoors spinach is relatively pest-free, but you can have issues with fungus gnats, spider mites, aphids, leaf miners, downy mildew, powdery mildew, and white rust.

VARIETIES & CULTIVARS 'Catalina', 'Winter Bloomsdale', 'America'. Newer varieties are slower to bolt, but may not produce as much seed.

HARVEST Harvest leaves by cutting at the base of the stem, from outside the central growing point, which allows new leaves to emerge from the center. The small young leaves will have the most flavor.

GROW THE SAME WAY French sorrel, blood sorrel

TIPS Spinach labeled *savoy* will have curly leaves; semi-savoy types will have leaves somewhat less curly. Flat-leaf types will be just that—flat. While attractive, savoy and semi-savoy are harder to clean than flat-leaf types.

Summer Squash

Cucurbita pepo

Squash is the collective name given to a number of species of plant in the genus *Cucurbita*, which includes pumpkins and winter squash, but the summer squashes are typically best suited to indoor growing. Plants produce large fruit in various sizes, shapes, and colors.

PHOTOPERIOD Day-neutral. Squash will flower under a variety of photoperiods, and you can grow them successfully with anywhere from 12 to 8 hours of light; 12 to 14 hours are adequate with high-intensity lighting. Use longer photoperiods with less intense lighting.

LIGHT REQUIREMENTS Bright light indoors, with direct full-sun exposure for 6 to 8 hours outdoors. Intense HO T5, CFL, LED, MH/CMH; or HPS combined with blue light from HO T5 or LEDs, spectrum-enhanced HPS, or Dual Arc.

PROPAGATION Direct seed only outdoors or into final container indoors. While many garden centers will offer transplants of squash plants and their relatives, these plants do not transplant well and growth is often stunted after transplanting. Seeds can germinate at temperatures between 60°F and 90°F (15–32°C), but germination is best achieved between 80°F and 90°F (26–32°C), as seeds can rot at cooler temperatures. Place a seed heat mat under containers when germinating.

TEMPERATURE Squash can handle warm temperatures, but perform best between 70°F and 75°F (21–24°C), so normal household temperatures are ideal. If you're growing squash inside a grow tent and temperatures are warmer during the day, be sure night temperatures are cooler.

SEASON Warm-season. Grow indoors year-round.

SPACE Standard squash plants are vines reaching 15 to 20 feet long, but dwarf bush squash will stay in the 3- to 5-foot range and are easy to keep in containers indoors. Choose a container at least 8 inches wide and about 12 inches deep.

GROWING MEDIUM Loose, well-draining, rich with organic matter, with a pH range of 5.5 to 7.5 (6.0 to 6.7 is ideal). Plants are suited to hydroponic, aquaponic, or aeroponic systems.

WATER Deep, consistent moisture is necessary, but overwatering can cause rot and problems with fruit development. Never water squash or zucchini plants on top of the foliage or you will encourage fungal diseases. Inconsistent watering can lead to blossom end rot, a deformity of the fruit caused by a disruption in calcium uptake.

FERTILIZATION Squash are heavy feeders. Add a balanced granular vegetable fertilizer to the container at seeding time. As vines begin to grow vigorously, you can feed weekly with a balanced organic liquid fertilizer, or side-dress with a granular several times until harvest. You can also apply a foliar spray of fish emulsion several weeks after seedlings emerge. Spray plants with a calcium phosphate solution to help prevent blossom end rot.

PESTS & DISEASES You should not have many pest or disease problems on indoor squash that receive adequate light, but whiteflies, fungus gnats, powdery mildew, and anthracnose can be a problem. Squash bugs and squash vine borers and fungal diseases are the most destructive pests outdoors.

VARIETIES & CULTIVARS 'Peter Pan', 'Gold Rush', 'Sunburst', 'Dwarf Summer Crookneck', 'Poquito'

HARVEST Harvest fruit while they are still tender and have reached the mature size described for each variety. Yellow squash left on the plant too long will get hard and inedible; green zucchini fruit, however, can grow to huge sizes while remaining soft and edible.

GROW THE SAME WAY Zucchini, pattypan squash, crookneck squash, straight-neck squash.

TIPS You will have to hand-pollinate indoor squash flowers. Squash and other cucurbits produce male and female flowers on the same plant. Male flowers typically emerge first, followed by female flowers, which have a swollen ovary (looks like a mini fruit) behind the flower.

Strawberry
Fragaria spp.

Strawberries are compact perennial plants that produce delicious fruits. Plants also produce attractive flowers, some very large and showy, making them perfect additions to edible landscapes. Their compact size makes them good indoor garden candidates, but be aware of different photoperiodic requirements among varieties.

PHOTOPERIOD Generally, June-bearing types are facultative short-day plants and will start flowering with shorter photoperiods of 10 to 11 hours. That's why you see strawberry plants blooming in late winter or very early in spring. Everbearing types are long-day plants and will start flowering closer to summer as days become longer than 12 hours. Temperature influences photoperiodism in strawberries, however. If temperatures are very cool, short-day strawberries can flower under longer days. If temperatures are very warm, long-day types can flower under shorter days. Day-neutral types, such as alpine strawberries, will flower and produce fruit year-round indoors, so you can grow them with longer photoperiods of 14 to 16 hours to boost production. Day-neutral types will be easiest indoors.

LIGHT REQUIREMENTS Bright light outdoors with full sun exposure for 6 to 8 hours outdoors. Plants benefit from a bit of shade in very hot outdoor climates. Indoors, intense HO T5, CFL, LEDs, MH/CMH; or HPS combined with blue light from HO T5 or LEDs, spectrum-enhanced HPS, or Dual Arc.

PROPAGATION Seed and plugs are available for Alpine and species strawberries. You can purchase bare-root plugs or transplants for most June or Everbearing types. New baby plants, called runners, will develop as the stems of the plant elongate. You can snip these off the elongated vines and place shallowly in potting mix, where they will take root and grow into new plants. Alpine strawberries don't make runners.

TEMPERATURE Normal indoor room temperatures are great for strawberries. Plants thrive between 68°F and 77°F (20–25°C). Strawberries can tolerate both warmer and cooler temperatures and still produce well.

SPACE Strawberries are great for small spaces and containers, and they grow 8 to 10 inches tall and 12 inches wide. A 6-inch pot per plant is adequate, or you can group plants in larger containers, hanging baskets, or grow towers.

GROWING MEDIUM Slightly acidic, rich with organic compost, with a pH level between 6.0 and 6.5. Plants are suited to hydroponic, aquaponic, and aeroponic systems.

WATER Even, consistent moisture that drains quickly. Do not overwater, especially in cooler temperatures, as crowns can rot quickly.

Strawberry
bare-root plant

FERTILIZATION Start with a rich soil, wait until plants begin developing baby fruit, and fertilize with a balanced fruit-and-vegetable granular fertilizer. Fertilize similarly every few months, or use a liquid feed monthly when plants are developing fruit.

VARIETIES & CULTIVARS Fraises des bois are some of the tastiest strawberries you will ever eat.

TIPS Boost production by hand-pollinating plants.

GROW THE SAME WAY Borage (day-neutral)

Thyme

Thymus vulgaris

Thyme is a classic culinary and aromatic herb with a pungent flavor. This low-growing bushy perennial comes in many varieties with varied flavors, growth habits, and foliage, and small pink to purple flowers. Many are quite ornamental in the landscape but may not be as well suited to culinary use.

PHOTOPERIOD Long-day. Thyme will produce more essential oils with longer days. You can grow thyme with 12 to 13 hours of light to delay flowering, or keep flowers pinched and harvest regularly under longer photoperiods.

LIGHT REQUIREMENTS Outdoors thyme needs to be in full sun with 6 to 8 hours of direct light. Indoors you can maintain healthy thyme plants using cool-spectrum HO T5, CFL, or LED lamps. You can also grow thyme under MH/CMH lamps for more intense production.

PROPAGATION Thyme can be slow and inconsistent when started from seed. You're better off buying transplants or taking cuttings. Tip cuttings, 2 to 3 inches long, typically root quickly and are an easy way to grow thyme. If you do seed thyme, sow seeds ¼ inch below the soil at temperatures around 70°F (21°C).

TEMPERATURE While thyme grows best at mild temperatures around 70°F (21°C), making it perfect for growing inside, it can tolerate a wide range of temperatures.

SEASON You can plant this cold- and heat-hardy perennial outdoors at any time during the growing season. Grow indoors year-round.

SPACE Thyme plants are compact, and you can typically maintain them longer in smaller containers than many other herbs. While they stretch only 6 to 12 inches tall, they do grow wide and can reach 12 to 24 inches in diameter. Choose 6-inch diameter or 2-gallon containers.

GROWING MEDIUM Loose, well-draining, low in organic matter, with a pH range of 6.0 to 8.0. Nutrient-rich soil can produce vigorous plants, but it will also reduce their fragrance and flavor.

WATER Thyme is a drought-tolerant plant. Water thyme thoroughly, but allow soil to dry completely before you water again. Thyme does not tolerate wet feet, and overwatering can quickly kill plants.

FERTILIZATION You won't need to fertilize thyme very much, but if you harvest regularly, feed plants with a half-strength solution of liquid humus, seaweed, or fish emulsion once per month.

PESTS & DISEASES You won't experience too many pest problems with thyme grown indoors, unless you overwater or humidity is too high. Most issues relate to fungal diseases such as gray mold and root rot. Fungus gnats can also be an issue.

HARVEST Snip shoots of thyme as you need them, fresh or dried. Store in a glass of water on the counter or in the refrigerator. Plants will get woody as they age, especially if you don't harvest frequently. Cut back severely if plants get woody, or start new cuttings.

VARIETIES & CULTIVARS English, French, lemon, orange, caraway, coconut

GROW THE SAME WAY French tarragon, lavender, perilla

Tomato

Lycospersicon spp.

One of the most popular of all vegetable crops (although botanically a fruit), tomatoes are a staple in many types of cuisine. Once you taste your own fresh, homegrown tomatoes, you will be hard pressed to eat the bland grocery-store variety. While popular, tomatoes are not the easiest crop to grow, as they are particular about temperature and water.

PHOTOPERIOD Day-neutral. Plants can be damaged by 20 hours or more of continuous light, so there is no benefit to lighting them any longer. Tomatoes typically produce well with 14 to 18 hours of light. If you are using less intense lights, grow with 16 to 20 hours of light. If you are using HID lamps that deliver more light, you can grow with shorter photoperiods of 12 to 14 hours.

LIGHT REQUIREMENTS Bright light indoors, with direct full-sun exposure for 6 to 8 hours outdoors. MH/CMH or intense HO T5, CFL, LEDs. HPS combined with blue light from HO T5 or LEDs, or spectrum-enhanced HPS, or Dual Arc. DLI: 20 to 30 mol m^{-2} d^{-1}.

PROPAGATION Sow seeds indoors and transplant outdoors after the last frost and when soil is warm, or sow seeds indoors and transplant into final growing container indoors. Sow seeds ¼ inch below the soil. Use a heat mat to keep soil temperature between 70°F and 80°F (21–26°C). Tomato seed germination is poor or nonexistent below 50°F (10°C) and above 95°F (35°C). You can also propagate tomatoes by cuttings and grafting.

TEMPERATURE Most new gardeners incorrectly assume that tomatoes are heat-tolerant. While tomatoes should be grown during the warm season, excessive heat can cause problems with blossom drop, poor fruit set, and fruit damage, as well as pest and disease issues that result from stressed plants. Tomatoes grow well with daytime temperatures between 70°F and 90°F (21–32°C) if night temperatures drop to between 55°F and 75°F (13–24°C). Night temperatures above 85°F (29°C) can cause heat delay, where plants don't set fruit. Conversely, plants will develop and fruit will ripen more slowly at the cooler end of the temperature range. Gardeners in cool climates often struggle to get tomatoes to ripen before temperatures get too cold. If the day and night temperature average hits 75°F to 78°F (24–26°C), that's perfect. Choose heat- or cold-tolerant varieties to match your growing climate.

SEASON Warm-season. Plant outdoors right after your last frost date, when soils begin to warm in spring (you'll benefit from planting earlier than your last frost date in hot climates), and again in midsummer in hot climates for fall production. Grow indoors year-round with appropriate temperatures.

SPACE Small to large. Determinate tomatoes are shorter and bushier, growing to 3 to 5 feet and putting on most of their fruit at the same time. Indeterminate

TOMATO

'Homeslice' tomato.

tomatoes are vining and grow large; some types can grow upward of 20 feet tall. Indeterminate types will continue flowering and fruiting if temperatures are ideal, but you will get less fruit at once. Determinate types are better suited to normal indoor growing spaces. Many new dwarf and mini-tomato plants are quite compact, making them perfect for container growing. For container size, plan on 2 gallons of pot volume per 12 inches of plant growth, and err on the side of larger containers. Fabric growing pots can lose too much moisture too quickly for tomato plants, especially in a warm environment. Stake plants, even the small ones, to provide support and minimize crowding.

GROWING MEDIUM Well-drained soil, rich with organic matter, with a pH range between 6.0 and 6.8. You can add coir to help maintain even soil moisture. Plants are suited to hydroponic, aquaponic, and aeroponic systems.

WATER Watering techniques and humidity will have a big impact on your tomato crops. The trick is to water evenly and consistently, but not too much. Keep water off foliage to discourage fungal diseases. Fluctuations between wet and dry conditions can cause problems with fruit and leaf development. Calcium deficiencies, which can result from inconsistent watering or overwatering, can lead to blossom end rot and leaf curl. High humidity may result in leaf fungal diseases, and the combination of hot heat and high humidity can disrupt fruit set.

FERTILIZATION Tomatoes need regular fertilization, especially in containers. Add a balanced tomato or vegetable fertilizer to the soil at planting time. Two weeks later, start feeding every other week with a 1:1:1 ratio liquid fertilizer. Once plants begin flowering and setting small fruit, switch to a formula with a 1:2:3 ratio and feed weekly. Many types of natural tomato fertilizers are available, and you can mix your own organic ingredients. You can spray tomatoes with a calcium phosphate solution as they mature to prevent blossom end rot or flower drop.

PESTS & DISEASES Indoors, your biggest issues will be fungal diseases such as leaf blight, powdery mildew, root rot, oedema, whiteflies, spider mites, fungus gnats. Aphids can be an issue on young plants.

VARIETIES & CULTIVARS Hybrids and heirlooms will have different characteristics. While heirlooms can have better flavor, most are large plants and can be more susceptible to diseases. There are thousands of varieties. Look for the determinate or indeterminate classification to determine size. 'Celebrity', 'Patio Hybrid', 'Lizzano', 'Nova', 'Little Napoli', 'Tiny Tim', 'Red Robin', 'Bush Early Girl', 'Bush Beefsteak', 'New Big Dwarf', 'Window Box Roma'.

TIPS Tomato root systems tend to be shallow but wide. Choose wide containers over tall, narrow ones. Tomatoes are wind-pollinated. You can boost fruit set by shaking the plants gently to move pollen around. Running fans in your growing area will also help pollination.

ORNAMENTAL PLANTS

I don't just grow plants; I live with them. Botanical cohabitation not only brings nature and beauty into our living spaces, but also engages us in a healthy caretaking experience. For people who cannot have furry companions, plants often take the place of traditional pets. Some plants grow just fine with bright ambient lighting from windows—no artificial light needed. If you're lucky enough to have many windows and a light, airy space, you can likely grow many low- and medium-light plants, as well as a few high-light foliage and tropical plants, without any supplemental lighting.

There are significant differences in light intensity as you move away from windows and between the rooms in your home. A plant that thrives in the windowsill may struggle just a few feet away. Sometimes even a bright window isn't enough. A plant is in trouble when it bends toward the lights and its growth slows.

Every home has unique light. The same type of African violet that blooms its head off in your friend's bright southern-exposure kitchen window may struggle in your north-facing living room. You will have to experiment with your space and find out what works best for you.

If you have a dark home but you want to keep large specimen plants in a low-light room, grow bonsai indoors, or force bloomers such as orchids, you will have much more success if you employ supplemental grow lighting.

If you start seriously collecting certain types of indoor plants or you want to enter specimens in flower shows and plant competitions, an indoor-light setup will be necessary to stimulate production and force flowering on a schedule.

This chapter covers a few popular categories of indoor ornamentals and reviews principles you can apply to many other plants with similar growing needs.

LEFT
A variety of ornamental plants can be grown indoors with enough light.

BELOW LEFT
Even low-light ferns will need supplemental grow lighting once you retreat to darker corners of your home.

BELOW MIDDLE
Mother-in-law's tongue (*Sansevieria* sp.) is a tough, easy-to-grow foliage plant that can survive in low indoor light levels.

BELOW RIGHT Dahlias growing under fluorescent lamps.

GESNERIADS

Gesneriads are a large family of more than 3000 tropical species that sport beautiful flowers and foliage. Many beloved blooming houseplants, such as African violets, are members of this group. Most gesneriads have similar growing needs and are perfect for indoor gardeners who don't have a lot of space but want to brighten up their home with flowers. Because this is such a large group, you can simply apply the guidelines for growing African violets to most other related plants in this family.

African violet
Saintpaulia spp.

Sporting succulent foliage and a wide array of flower colors, African violets are the perfect compact houseplant for forgetful waterers.

PHOTOPERIOD Day-neutral. Do not light for longer than 16 hours, however, as this can suppress flowering.

LIGHT REQUIREMENTS Bright to medium. If you have windows that get bright light but are shielded from direct sun during the hottest part of the day, you can usually grow African violets in your windowsill without supplemental lighting. Hot direct sunlight can scorch leaves. If plants are stretching or are not blooming, use HO T5, CFL, or full-spectrum LEDs to provide additional light. Place lamps about 12 to 18 inches above larger plants but only about 12 inches above miniature varieties. Leave your lamps on 12 to 14 hours per day. If you're growing African violets or other gesneriads using only artificial light, you may get better results lighting for 14 to 16 hours. You can also rotate plants to keep their growth habit more balanced.

PROPAGATION Crown divisions and leaf cuttings. African violets, as well as many other gesneriads, are easy to propagate from cuttings. You also can grow from seed, which germinates in 8 to 14 days. Temperature for the soil should be between 65°F and 75°F (18–24°C). However, if you're looking to clone a specific cultivar, stick to cuttings. Remove a healthy leaf with the petiole or leaf stem by cutting or snapping it off neatly at the stem of the plant. Trim the petiole to 1½ inches for best results. Insert the leaf stem into a hole in the growing medium, then water it thoroughly.

TEMPERATURE Ideal growing temperatures fall between 65°F and 75°F (18–24°C), but plants can perform well at a wider range of 60°F to 80°F (15–26°C). They are generally comfortable at the same indoor temperatures you are.

SPACE African violets and other gesneriads are found in a range of sizes, but most are compact and good for small spaces. Miniature varieties can be maintained in 2- to 3-inch diameter containers, while standard specimens can be maintained in 4- to 6-inch containers. Other species of gesneriads grow larger, especially vining types, and can fill a 10- to 12-inch container.

GROWING MEDIUM Lightweight, well-draining, heavy on coir or peat. African violets don't tolerate wet feet, and heavier soils with too much compost will not allow them to dry out enough fast enough between waterings. You can also add expanded shale or perlite to loosen the mix. You can repot African violets annually.

Begonias prefer to dry between waterings.

WATER Plants need to dry between waterings and can rot quickly if you leave them too wet too frequently. If you are a forgetful waterer, you'll do great with African violets. Make sure the soil is dry to the touch before watering, and always keep water off the foliage, as it can cause spots on the leaves as well as fungal issues.

FERTILIZATION A balanced formula is best. Use at each watering, following the manufacturer's instructions. Avoid bloom-boosting fertilizers with excessive phosphorus, as they can cause foliage to appear scorched.

PESTS & DISEASES Thrips and mealy bugs are the most common indoor pests of African violets and other gesneriads, but stressed plants can also fall victim to spider mites. More commonly, your plants may succumb to soil-borne fungal diseases or powdery mildew. You can manage both with good moisture and humidity control.

VARIETIES & CULTIVARS There are more than 20 wild species of African violet, and countless varieties and hybrids.

TIPS Do not let your violets get too large or develop extra crowns, or small sucker plants. Plants will look and bloom best if you keep them to five or six rows of leaves. Divide out new crowns and make new plants. The exception is trailing varieties; you can allow them to develop extra crowns without loss of plant vitality.

GROW THE SAME WAY Gesneriads such as *Achimenes*, *Aeschynanthus*, *Columnea*, *Episcia*, *Kohleria*, *Nematanthus*, *Sinningia*, and related groups such as *Begonia*, *Hoya*, and *Iochroma*

BONSAI

If you live in a home without outdoor garden space, you may be inclined to buy and train bonsai trees so you can bring some nature inside. There is nothing more tempting than a beautifully sculpted bonsai tree, which are often sold as "easy" indoor plants for novice gardeners. If you have at some point succumbed to such a temptation, only to find your pricey bonsai tree dropping leaves a few weeks after it has been indoors, you're not alone. Most species of plants trained as bonsai are meant to be maintained outdoors. But that's not how bonsai typically are sold, and most beginner bonsai-ers end up learning this lesson the hard way.

An elm tree grown as a bonsai specimen.

The Right Light

A few environmental factors will influence your success with indoor bonsai. Light, of course, is a big one. Most woody plant species that are trained as bonsai need more light than they would typically get with ambient indoor lighting. They need the equivalent of partial-to-full sun exposure that they would normally receive outdoors. Even a bright, sunny window won't always do the trick. Most often, you'll need to combine a sunny window with a supplemental grow light. Twelve hours of light from HO T5, CFL, or full-spectrum LED lamp, in conjunction with bright ambient light, is recommended. If your bonsai won't receive much ambient light at all, then you may need a more intense grow lamp.

Certain tropical plants can also serve well as trained bonsai. These tropicals will be better suited to living indoors year-round, but you will still need supplemental light for many types unless they are low-light plants.

The Right Temperature

Temperature and dormancy will also determine your success with indoor bonsai. Most temperate species require a dormancy or resting period. This process typically requires a drop in temperature that does not occur naturally indoors. You may need to transition your bonsai outdoors for a period in the fall to trigger a dormancy. Alternately, you can try a cool basement, balcony, or unheated porch. Deciduous species of bonsai will drop their leaves, just as they would in fall and winter outdoors. Give them some time in a dormant period before transitioning them back indoors, which will trigger them to begin growing and leafing again. If you anticipate a severe drop in or sudden, deep freezing temperatures outdoors, you'll need to cover and protect your bonsai or bring them indoors temporarily; small plants in shallow containers aren't insulated by warmth in soil, as your landscape plants are.

A bonsai grown under a spotlight LED.

Watering your bonsai properly is another skill you will acquire through practice. Bonsai are grown in very shallow containers with little soil volume to hold moisture. A juniper bonsai, for example, can quickly dry out and die before you ever know it has happened; often the dead foliage will still hold a pale green color, tricking you into thinking the plant is still alive. Set your bonsai container in a shallow tray of water and allow it to soak up moisture through the drainage holes in the pot. You may need to do this once or several times per week, depending on the type of plant you have.

There is no set schedule for repotting bonsai. Rather, you'll need to inspect the root system of each plant every spring by gently lifting the plant out of the pot. If the roots have started to circle around the pot, it's time to remove the plant, trim some of the roots, and repot. If all the roots are still in soil and haven't begun to circle the pot, leave it alone and wait until the following spring to check it again.

Bonsai plants don't have much access to soil nutrients, so you must fertilize them on a regular schedule. From early spring through summer, you can feed plants weekly with a natural humus-based liquid fertilizer. If you use a higher-nitrogen synthetic fertilizer, you can dial that back to every other week or monthly. Reduce fertilization in the fall for temperate species and give tropicals a break in winter as well.

Learning to prune and train your bonsai will take many years, but it's a meditative and rewarding practice. The trick is to keep your plants healthy and vigorous over the long term so you can create artful forms over time. You can train many plants into a bonsai specimen, but some are better suited to the training and small containers than others.

Popular Species for Bonsai

DECIDUOUS BONSAI	EVERGREEN BONSAI	TROPICAL BONSAI
Baobab—*Adansonia digitata*	Azalea—*Rhododendron* spp.	Bodhi tree—*Ficus religiosa*
Cherry—*Prunus* spp.	Boxwood—*Buxus* spp.	Bougainvillea—*Bougainvillea* spp.
Crabapple—*Malus* spp.	Cedar—*Cedrus* spp.	Desert rose—*Adenium obesum*
Chinese elm—*Ulmus parvifolia*	Dwarf Japanese holly—*Ilex crenata*	Dwarf schefflera—*Schefflera arboricola*
Cotoneaster—*Cotoneaster horizontalis*	Juniper—*Juniperus* spp.	Fukien tea tree—*Carmona retusa*
Dwarf pomegranate—*Punica granatum*	Olive—*Olea europaea*	Jade bonsai—*Crassula ovata*
Japanese elm—*Zelkova serrata*	Pine—*Pinus* spp.	Money tree—*Pachira aquatica*
Japanese maple—*Acer palmatum*	Snow rose—*Serissa foetida*	Panda ficus—*Ficus americana*
Trident maple—*Acer buergerianum*	Yew—*Taxus* spp.	Weeping fig—*Ficus benjamina*

CARNIVOROUS PLANTS

Their graceful and alien-like appearance, plus their compact size, makes carnivorous plants the perfect specimens for indoor collectors. While you're likely familiar with the most well-known insect eater, the Venus flytrap, there are more than 670 species from which to choose.

Carnivorous plants grow in boggy, acidic soils that are primarily composed of peat and have few of the nutrients and minerals plants need to survive. Nitrogen, the most important plant macronutrient, is very quickly leached from wet soils. In order to replace the nitrates, carnivorous plants evolved to digest insects.

The Right Light

While you could successfully grow a few species of carnivorous plants in a bright window, most need significantly more light and humidity. In fact, most carnivorous plants grow naturally in full-sun locales, so you must provide intense light indoors for your creature-capturing plants. Some tropical species do best when planted in a high-humidity terrarium. If you want to plant different species in the same terrarium, remember to group together those with similar environmental needs.

American, or purple, pitcher plants (*Sarracenia* sp.) growing alongside sundews in a glass conservatory.

You can also grow carnivorous plants on standard shelving with two to four HO T5 fluorescent tubes, or several CFLs or LED bars, suspended about 12 inches above your plants. Light your carnivorous plants for 12 to 16 hours per day, depending on your light source and the ambient light available in your living space.

If you plan to get serious about your carnivorous plant addiction and feel the need to breed, you can graduate to HID lighting in the form of MH/CMH lamps with cooling climate control. As plants mature, be sure to increase the distance between your plant and lamp more significantly than you would for fluorescent/LED lighting.

The Right Temperature

Many carnivorous plants are cold-hardy perennials that require a winter dormancy to thrive. Shortening days and cooling temperatures trigger them to go dormant for anywhere from 10 weeks to 3 months. Without dormancy, your plants will decline. This can be a little tricky to achieve indoors, but if you're growing your plants under lights, reduce the photoperiod to 10 hours in the fall and then increase it again in spring. Cooler temperatures will also help, so try moving plants to a cooler spot in your home in the winter or turning down the thermostat. Some growers even resort to unpotting their specimens and putting them in the refrigerator. If you're growing species native to the eastern coast of the United States, they most likely have a dormancy requirement. Tropical species that do not require dormancy will be easier to grow indoors year-round. All carnivorous plants require high humidity.

Avoid standard potting soil or compost as your growing medium. Use only sphagnum peat moss as your substrate. To loosen the peat, you can supplement with perlite, lava rock, expanded shale clay, or triple-rinsed playground sand (not builder's sand).

You should avoid tap water or bottled spring water, which contain harmful salts and minerals, for carnivorous plants. Use dechlorinated distilled water, water that has gone through a water softener, or collected rainwater.

Pests and diseases won't be much of a problem for your carnivorous plants. Snails and slugs, as well as larger animals, can be an issue in the outdoor garden.

The unique flowers of an African pitcher plant.

Feeding

Carnivorous plants are meat-eaters; they need to consume a few insects now and then to remain vigorous. Each type of trap will take about a week to digest a bug. If it's an active trap, such as the Venus flytrap, it will close when fed and reopen after it has finished digesting. You likely do not have enough insects flying about your home or inside a terrarium to supplement your plant's diet. If you provide the insects, make sure you feed them live to your plants. If you can't do that, buy dehydrated blood worms (DBW) at your local pet store. Rehydrate DBW with water, which will make a sort of meat paste. Plan to feed your plants seasonally, four times per year, with the equivalent of three to four bugs or drops of DBW paste, per feeding and per trap.

Easy Carnivorous Plants to Grow Indoors

Cape sundew: *Drosera capensis* can be grown in a very bright window or under lights. Easy to grow with clean water.

Tropical sundews: *Drosera aliciae*, *D. adelae*, *D. binata*, and *D. spatulata* are all lovely, bright-light specimens that can be grown together under the same conditions to create a beautiful terrarium planting. In fact, it is best to grow tropical sundews only inside terrariums, otherwise they will dry out too quickly.

Butterworts: *Pinguicula* are tiny medium-light plants with shallow root systems. If you don't have the brightest light available, you might try butterworts. These little plants do have a dormancy, but rather than going to sleep in winter, they develop swollen succulent leaves that store water instead of catching insects. All you have to do is stop watering in winter.

Drosera intermedia has a striking form.

Asian pitcher plants: *Nepenthes* can be grown in hanging baskets in bright light. Look for plants labeled as highland types for easy growing, or take on more of a challenge with lowland types, which will fare better in a terrarium.

Australian pitcher plant: *Cephalotus follicularis* are small-growing plants that can tolerate medium- to high-light conditions with a wide swing in temperature, but they are happiest in normal indoor temperatures. This species does best inside a terrarium, but make sure to provide good drainage. It dislikes being waterlogged.

Bladderworts: *Utricularia* species are some of the tiniest and sweetest carnivorous plants you can grow. Their size and tolerance for waterlogged soils makes them perfect for miniature indoor gardening, as you can plant terrestrial (ground) types in just about any kind of container—even without a drainage hole. They need bright light and plenty of water. You can also plant aquatic types in aquariums.

FAR LEFT *Nepenthes (viking × rafflesiana) × (ampullaria × northiana)* has unusual burgundy coloring.

Pinguicula primuliflora produces primula-like lavender flowers.

CARNIVOROUS PLANTS

Orchids, ferns, and herbs growing in my kitchen.

ABOVE RIGHT
A purple oxalis growing as a houseplant under fluorescent T5 counter lights.

RIGHT You can light specimen plants with spotlight grow lamps.

FOLIAGE PLANTS

You can grow many foliage plants indoors without supplemental light, and there are plenty of books and resources available to teach you how to cultivate specific types. If your space does not provide enough ambient light, a grow-light system will fill the gap. You can grow small-foliage plants on counters or shelves under cool-spectrum HO T5 fluorescents or on tables and desktops using full-spectrum LEDs.

Large floor-plant specimens can be a bit trickier, as they often require more light than the living room corner offers. If you want to avoid hanging a big metal fixture for a large cool-spectrum CFL bulb—what you need for, say, a large fiddle-leaf fig or palm tree if it's in a dimly lit location—you will have to work with several smaller CFL spotlights. Place lamps within 6 to 12 inches of your plants. Leave them on for 16 to 18 hours per day if plants receive little to no light from windows. If there is partial ambient light, 12 to 14 hours are adequate.

Try growing foliage plants in one part of your home or office. If they are just not thriving, move them closer to an ambient light source. If that still doesn't do the trick, add artificial light or switch to a lower-light species.

The fiddle-leaf fig is one example of a beloved houseplant that can give you some trouble without supplemental lighting. You can use its growing guidelines to help you care for many other common types of indoor foliage plants.

Fiddle-leaf fig

Ficus lyrata

There is no cooler houseplant than the fiddle-leaf fig. Its sculptural good looks enhance any stylish living space. But this member of the fig family can be challenging to care for. Plants need just the right amount of light, the right amount of water, and the right amount of humidity to thrive.

PHOTOPERIOD Day-neutral. You're not looking to push blooms, so the duration of light ensures enough light accumulation through the day so plants don't drop their leaves. Start by lighting the plants for 12 hours a day. If they are still not thriving, extend the photoperiod to 16 to 18 hours.

LIGHT REQUIREMENTS Medium- to high-light requirements year-round. Bright, diffused light, with just a little direct sun (but not too much). Despite all my north-facing windows, there is no room in my home bright enough for my fiddle-leaf figs to thrive without supplemental light. Direct east-facing windows are preferable so plants get bright morning sun with a bit of direct sun, but are shielded from direct hot-afternoon sun. Use cool-spectrum HO T5,

CFL, or LED lamps outfitted with standard sockets in spotlights or directional lamps. Several lamps may be necessary, depending on your indoor light levels.

PROPAGATION Cuttings, air-layering.

TEMPERATURE The ideal growing temperature range is between 60°F and 75°F (15–24°C) consistently. Fiddle-leaf fig and its relatives do not like change and can drop leaves if temperatures swing wildly or grow suddenly cold or warm.

SEASON A tropical that can be grown indoors year-round or outdoors year-round in tropical climates. In temperate climates you can move plants outdoors in summer, but you must bring them inside as temperatures cool. Remember, temperature fluctuations can cause leaf drop.

SPACE Fiddle-leaf figs are trees. Make sure to provide a container that has at least a 5-gallon volume, but you may need to increase the size as your plant matures. You can repot every two years in spring or summer.

GROWING MEDIUM Loose, well-draining, with some organic matter. If you tend to be a forgetful waterer, you can incorporate some coir into the potting mix, which will help hold more even moisture over time.

WATER These plants need to be watered once every week. If yours is 5 to 7 feet tall, give it about a quart (1 liter) of water. If yours is larger, it will need about a gallon (4 liters). Make sure the soil always remains slightly moist. Fiddle-leaf figs wilt when they need water but will perk up quickly if you provide it soon enough.

Fiddle-leaf fig

FERTILIZATION Fertilize with a balanced houseplant fertilizer or natural liquid fertilizer and add ½ to 1 inch of organic compost or humus to the top of the container once each spring. Fertilize monthly through summer months. Stop fertilizing midfall through winter.

PESTS & DISEASES Leaf spots from fungal diseases are a common problem caused by overwatering, poor water drainage, and stagnant airflow. Spider mites can be an issue, especially on plants not grown in ideal conditions. Brown spots from sun scorch can also be a problem if plants are exposed to hot direct sunlight. Once leaves develop large brown spots, they will not recover. Cut off the leaves and dispose of them, taking care not to get the irritating sap on your skin.

VARIETIES 'Suncoast', 'Compacta'

TIPS Pay attention when you're buying plants. Too often, medium- to high-light plants are kept in conditions that are far too dark for far too long, and they will decline and present challenges once you get them home. Be sure to select specimens that are clearly housed in a bright location and look full and healthy.

GROW THE SAME WAY Rubber tree (*Ficus elastica*), *Monstera* spp., *Philodendron* spp., parlor palm (*Neanthe bella*), *Dracaena* spp., to name a few. There are many other tropical foliage plants you'll grow under similar conditions that benefit from supplemental light.

ORCHIDS

With 25,000 to 30,000 known species of orchids from different regions all over the world, you can imagine the broad variety of environmental and growing needs. This section groups some of the most popular orchids into primary care categories to help you learn the basics.

Many orchids are finicky in their requirements. As a beginner, you may find certain types of orchids challenging to grow or reflower indoors, while others will be relatively easy. Conditions in your home are the main determining factor. Don't get discouraged if you aren't successful with an orchid on your first try. I have killed more orchids than I can count, but in doing so I have also figured out exactly which ones I'm good at growing. While orchids used to be significantly less available and far more expensive, today's market offers a bounty of options at very reasonable prices. You can start to experiment with easier and less expensive types or jump right into the deep end of the pool with pricier ones that require more care. Experimentation is always part of the fun.

The Right Light

Orchids have three general categories of light needs. The color of their leaves will tell you if they are getting too much or too little light. Normal, healthy leaves should generally be a light, bright green. Leaves that become very dark generally aren't getting enough light, while a pale yellow-green or red tint indicates plants are getting too much light. (This can be counterintuitive for most orchid newbies, as paler leaf color often signals diminished health in other types of plants.) There will, of course, be some variation on leaf color depending on the species. Plants that receive too much hot direct sunlight or are placed too close to a light source can scorch. If this occurs, be sure to remove any damaged leaves.

A display of a variety of orchid species in bloom.

Miltonia orchid sports impressive blooms and a nice fragrance.

HO T5, CFL, and full-spectrum LED lamps are most commonly used to grow orchids. You may need more lamps or more intense lamps to provide enough light. You can, of course, alter the spectrum of light to trigger flowering in fall and winter. If you're serious about your collection and you want to boost growth and flowering, you can use HID lamps with success—especially if you're growing your orchids where they won't receive any other supplemental light, such as in a basement, a closet, or a grow tent. You will need to be mindful of temperature control with HID lamps. CMH lamps are a good HID choice for orchids, with temperature-control measures for cooler-temperature orchids.

Orchid Photoperiods

Many orchids are day-neutral. They flower in response to cooling temperatures in fall and winter. That means you can typically light groups of most orchids from 14 to 16 hours a day, depending on how much ambient light they receive.

However, a few orchid species require short-day photoperiods in conjunction with cool temperatures to flower, such as some species of *Cattleya, Dendrobium*, and potentially some species of *Phalaenopsis*. For these types, you'll need to combine cooler temperatures and shorter daylengths from midfall through winter, then lengthen your lighting period again in spring.

Other orchid types, such *Zygopetalum*, may not require short days to bloom, but can flower faster and better when grown under short days combined with cool temperatures. That makes them facultative short-day plants. Some literature claims shorter photoperiods can trigger *Phalaenopsis* to flower faster, while other sources counter that cool temperatures are doing the trick. Home gardeners can simply treat *Phalaenopsis* as a day-neutral orchid. If you're serious about growing and forcing orchid blooms, research your desired species and experiment with light and temperature to see what garners the best results.

General Care

In general, most orchids prefer high humidity with a very well-drained and aerated potting mix. Many common orchids are epiphytes (air plants) that can't survive in

Catteleya orchids are large specimens that love a cool spot.

Zygopetalums are one of my favorite orchids; the flowers have a wonderful fragrance.

a wet, heavy potting mix. There are preformulated orchid potting mixes that are appropriate for many types of orchids. Some high-moisture orchids will benefit from additional sphagnum moss added to the mix, while more terrestrial-type orchids favor some compost blended in. It's best to repot your orchid every two years to replenish organic matter. If the roots still fit, use the same size pot. If the root system has gotten larger, choose a container that just fits the root system's size. Pots that are larger than the root system can make water management more difficult and reduce flowering.

Water

Most orchids do well when watered once per week. Ideally, you should use tepid dechlorinated water. Let water run slowly over the potting mix for a minute or so. You can also set orchid pots in water until it soaks through the growing medium. Then remove and let all excess water drain out of the pot and saucer. Wetting the growing medium—as opposed to directly watering roots like you might with terrestrial-type plants—increases the humidity around the root zone and leaves. Epiphytic types with aerial roots will appreciate a regular misting of foliage and roots versus heavy watering at the root zone.

You may have heard that you should water your orchids using ice cubes because the slow melt will deliver the right amount of water over the right amount of time. I recommend avoiding this approach, which is meant to mitigate potential overwatering in the heavy moss and nonporous containers in which you often

Vanda orchids require high humidity in order to thrive.

find orchids for sale. If you plant your orchid in a container that provides good drainage and aeration and use a potting mix formulated for orchids, you'll be better off watering as recommended here.

Some orchids, such as *Dendrobium* species, can grow happily at very warm summer temperatures with regular moisture, but if they don't experience a much cooler and very dry winter, they may not bloom. You'll have to put the brakes on watering for several months. Other orchids will still need consistent watering through winter to look their best. Be sure to learn the specific seasonal requirements of each type of orchid you grow so you can properly tailor your maintenance regimen through the year.

Fertilizer

While different types of orchids will require different fertilization regimens, most are not heavy feeders. Fertilizing plants with an orchid fertilizer or an organic liquid feed once per month during the active growing season is generally adequate. Most orchids can get by with less watering and humidity, as well as reduced fertilization, during winter months. Consider winter their resting period. I do not recommend trying to overstress your orchids to get them to flower. You will just end up shortening their life spans. Keep watering and feeding your plants year-round, except those types that need a dry dormancy in winter.

When choosing which orchids to grow, consider your growing space and the kind of light and temperatures you can provide. Orchids are adaptable to a degree. Many can still perform well with some variation outside of their normal light and temperature needs.

Keep in mind that the recommended temperature ranges here are for winter months (summer if you are south of the equator). Most orchids can tolerate warmer spring and summer temperatures that fall outside these ranges. There will also be some overlap between temperature ranges for different orchid types, and modern hybrids may differ in their needs. If you grow your orchid consistently outside its ideal general temperature range, however, or if it does not experience a meaningful temperature drop at night, it may become stressed and fall into decline.

There is a long list of diseases and pests that can affect your orchids. Most often, some form of bacterial or fungal rot attacks your plants. These diseases can cause spots on the leaves or rot plants at the root level. Good water and humidity management is the most efficient way to prevent these problems. You may also have issues with mealy bugs or scale. Treat them with an insecticidal soap (rather than an oil) and remove them from the plant using cotton swabs or cloths.

An orchid planted in a porous pot with a loose orchid bark mix.

Dendrobium orchids typically require high light levels to thrive.

ABOVE RIGHT Never allow dracula orchids to dry out.

ORCHID TEMPERATURE CATEGORIES

WARM Daytime temperatures between 80°F and 90°F (26–32°C) and night temperatures between 65°F and 70°F (18–21°C). Warm-category orchids often also need high humidity: *Phalaenopsis*, *Vanda*, warm-temperature *Dendrobium*.

INTERMEDIATE Daytime temperatures between 70°F and 80°F (21–26°C) and night temperatures between 55°F and 65°F (13–18°C). These orchids are typically the easiest group for indoor gardeners because they grow in temperatures most similar to home temperature settings: *Paphiopedilum*, *Epidendrum*, *Cattleya*, *Miltoniopsis*, and intermediate-temperature *Oncidium* types.

COOL Daytime temperatures between 60°F and 70°F (15–21°C) and night temperatures between 50°F and 55°F (10–13°C): *Cymbidium*, *Dendrobium*, *Masdevallia*, cool-temperature *Oncidium*. For cool-temperature orchids, a lighted basement, an unheated closed-in porch, or a sunroom provide a good winter environment. A particularly drafty side of your home near a window will also serve.

Moth orchid
Phalaenopsis spp.

Phalaenopsis are the most commonly sold orchids these days. You will find them en masse at most grocery stores and garden centers. There are many species and flower colors available, and blooms can last three to four months. They are relatively easy to grow and rebloom, as long as you meet their basic light, temperature, and watering needs.

PHOTOPERIOD Most species and cultivars are day-neutral. Grow with 14 to 16 hours of light. Plants may be considered quantitative short-day, as short photoperiods of 12 hours (long nights) in late fall and winter may speed up flowering a bit. It is primarily cool temperatures, however, that trigger *Phalaenopsis* to flower.

LIGHT REQUIREMENTS Low to medium. Increase light in fall and winter to encourage flowering. An east- or west-facing window is required if using natural light, but hot direct sun or artificial light that is too close to plants can scorch the foliage. Leaves should be upright. If they get large and floppy, plants are not getting enough light. Plants grow well under supplemental lighting with fluorescent, HO T5, CFL, and full-spectrum LED. Ambitious collections may use MH/CMH or extend photoperiod with HPS.

PROPAGATION Division. Plants will produce new plantlets that you can divide and pot up.

TEMPERATURE *Phalaenopsis* grows best in warm temperatures between 66°F and 86°F (19–30°C) during the day. Ideal night temperatures fall between 60°F and 66°F (15–19°C). If your plants are vigorous and healthy but not blooming, drop the temperature by 8 to 10 degrees Fahrenheit or move the plant to a cooler spot for several weeks. Spikes should develop.

SEASON New spikes will begin to naturally form in late fall and early winter, but you can induce flowering at other times of year by reducing temperature. You can also trigger new bloom spikes by cutting off an expired spike at the next viable node on the stem.

SPACE Small-space plant. Orchid pots in the 4- to 8-inch-diameter range are typically adequate. You can grow mini cultivars in 2-inch pots.

GROWING MEDIUM Use a coarse bark-based orchid mix for epiphytes. Repot every one to two years.

WATER Water once per week with tepid water. Run water slowly over pot and roots for several minutes. Do not let pots sit in water that has collected in a tray. You can cut back a bit on watering during winter months. Misting plants in the growing season is helpful, but always wipe excess moisture from the foliage to prevent disease.

FERTILIZATION When you water weekly during spring, summer, and fall months, fertilize for three consecutive weeks and skip the fertilizer every fourth week. Reduce fertilization to once per month in winter. Use a water-soluble orchid fertilizer or liquid humus.

Lady slipper orchid

Paphiopedilum spp.

Lady slipper orchids are one of the terrestrial orchid types commonly grown indoors. Plants naturally flower from about November through March, but given the right conditions flowers may appear at other times throughout the year. Lady slippers, some of my favorite orchids, are fairly easy to grow. Multifloral types will produce more than one flower per stem in succession. Complex types produce only one flower per stem.

PHOTOPERIOD Day-neutral. Grow with 14 to 16 hours of light. Plants may be considered facultative short-day, as short photoperiods of 12 hours (long nights) in late fall and winter can speed up flowering a bit. However, it's primarily cool temperatures that encourage *Paphiopedilum* to flower.

LIGHT REQUIREMENTS Low to medium. Increase light in fall and winter in conjunction with a temperature drop to promote flowering. An east- or west-facing window is required if using natural light, but hot direct sun or artificial light that is too close to plants can scorch the foliage. For best performance, grow under supplemental lighting with HO T5, CFL, full-spectrum LED. Ambitious collectors may use MH/CMH or extend photoperiod with HPS with climate control.

PROPAGATION Division. Plants will produce new plantlets that you can divide and pot up.

TEMPERATURE There are two groups of *Paphiopedilum* plain, green-leaf types that prefer slightly cooler daytime temperatures of up to 77°F (25°C) and nighttime temperatures between 50°F and 55°F (10–13°C), and mottled-leaf types that like it a bit warmer, with daytime temperatures up to 86°F (30°C) and nighttime temperatures around 65°F (18°C). The mottled-leaf types are usually easier to grow indoors under normal household temperatures.

SPACE Small-space plant. Containers in the 4- to 8-inch-diameter range are typically adequate. This is one of a few orchid groups for which you should not use orchid pots with slits; instead, use a traditional container with a drainage hole. This will help you maintain better moisture levels.

GROWING MEDIUM Use a bark-based orchid mix, then add some sphagnum or peat moss to the mixture to increase moisture retention.

WATER Keep the potting mix moist, but never let plants sit in water. Allow plants to dry just a bit between waterings, once or twice per week. Do not mist.

FERTILIZATION Fertilize twice per month through spring, summer, and fall with a water-soluble orchid fertilizer or liquid humus. Reduce or eliminate fertilizer in winter months.

Boat orchid

Cymbidium spp.

Their large plant size, massive flower spikes, and long-lasting blooms make *Cymbidium* one of the most popular and long-cultivated groups of orchids. Plants are fast growing, and established plants can produce many spikes of flowers over winter months. Their thick, waxy blooms are well suited to flower arrangements, bouquets, and corsages.

PHOTOPERIOD Day-neutral. Grow with 14 to 16 hours of light year-round. There is no benefit to changing the photoperiod in an attempt to trigger flowering.

LIGHT REQUIREMENTS High. Some *Cymbidium* prefer direct sun, so you'll need bright light. Increase light in fall and winter in conjunction with a temperature drop to encourage flowering. A south- or west-facing window is required if using natural light, but plants generally perform better when the light is delivered above the plant, so you may have more success with overhead supplemental light. Use HO T5, CFL, full-spectrum LEDs, or MH/CMH for more advanced production needs; extend photoperiod with HPS with climate control.

PROPAGATION Division. Plants will produce new bulbs that you can divide and pot up.

TEMPERATURE *Cymbidium* needs cooler growing conditions than most tropical orchids. A drop in night temperature in fall and winter is especially important to triggering *Cymbidium* to bloom. In spring and summer, always grow plants at temperatures below 86°F (30°C). If you live in a hot climate, bring *Cymbidium* indoors for the warm season. In order to encourage flower spikes, you need to expose plants to temperatures in the 50°F to 57°F (10–14°C) range. You can do so by setting plants outside in mid- to late summer through fall, depending on your climate, to expose them to cooling temperatures. Once temperatures reach freezing and flowers have begun to open, bring plants indoors and keep in a cool place in your home.

SPACE Large, but there are also miniature types. Contrary to how you pot other types of orchids in tight containers, *Cymbidium* should be given large containers (bigger than the root ball) to allow for new bulb formation.

GROWING MEDIUM Use a medium-grade orchid mix and blend in some sphagnum or peat moss to help keep moisture consistent. Repot plants in spring every two to four years.

WATER Moderate water needs from spring through summer. Water plants weekly from above and never allow plants to sit in water. You can reduce watering to once every two weeks in winter, but do not allow bulbs to shrivel because they are too dry.

FERTILIZATION Use a half-strength solution of liquid orchid fertilizer or a liquid humus fertilizer every three weeks in spring. You can switch to a fertilizer with more potassium in summer months, then go back to the half-strength formula in fall and winter.

GROW THE SAME WAY Cool-temperature *Dendrobium*, *D. nobile*, *D. speciosum*, and cool-temperature *Oncidium*

SUCCULENTS

While succulents are all the rage among gardeners, many don't fare well indoors over the long term. Most succulents and cacti prefer high-light conditions. Remember that while all cacti are succulents, not all succulents are cacti. It's important to know the difference when choosing indoor specimens. While some succulents can tolerate lower levels of indoor light, most cacti need a lot of light to thrive. If you are tired of growing succulents or cacti indoors only to find they get spindly and eventually sputter out, low light levels are most likely to blame—and too much water.

The Right Light

Most of my windows are on the north side of my home. While this orientation is fine for plants that don't need intense light or afternoon sun, such as African

CLOCKWISE
FROM TOP LEFT

Hens and chicks, *Sempervivum* species, are popular container plants.

This ruffled echeveria will need bright light to survive.

Cacti are sculptural specimens that need bright light.

CLOCKWISE FROM TOP LEFT

Mother-in-law's tongue, *Sansevieria* 'Fernwood', is compact and has thin leaves.

Zebra plant, *Haworthia*, is a succulent that tolerates lower light levels.

Aloe vera is the classic species, but many other colorful types can tolerate indoor conditions.

Ponytail palm, *Beaucarnea recurvata*, looks more like a tropical plant but tolerates lower light and limited water.

violets, it is just not bright enough for most sun-loving succulents, and they decline even when placed directly in the windowsill. Several types of shade-tolerant succulents can thrive in a bright window without grow lighting.

Crassulas—a group of succulents that includes aeoniums, sempervivum, jade plants, kalanchoes, and adromischus—can handle medium light levels that you can provide with bright windows indoors. If you have east-, south-, or west-facing windows that channel at least several hours of direct sunlight, you will fare better with these succulents as well as higher-light plants such as echeverias. But they will go downhill if your window isn't bright enough or if you move them into lower-light conditions.

Most succulents are happiest outdoors with bright light during the warm growing season. Place them in a location where they will get direct sun all morning or through the early afternoon, with a bit of dappled shade in the late afternoon. Sun-loving cacti and succulent agave can be in full sun all day. However, each type of succulent tolerates certain temperature ranges, and not all are frost-hardy, especially in containers. If you live in a climate with freezing winter temperatures, you will need to bring many of your succulents indoors for the winter.

By employing supplemental lighting indoors, you can keep a healthy collection of better-looking succulents year-round. You can also grow succulents under lights year-round to keep them healthy, but set them out in your home in lower light conditions for short periods of time to enjoy.

HO T5, CFL, and full-spectrum LED lamps will be most useful if you are growing succulents in your open living space. If you plan to grow your succulent collection tucked away in an office, a grow closet, or another out-of-the-way space, you might choose an LED with a more limited spectrum (pink-colored light). If you're using HO T5 tubes, succulents can benefit from mixing in a red-spectrum lamp with your cool-spectrum lamps. Typically lamps should be placed close to plants, about 6 to 12 inches from the top. Most succulents are accustomed to 6 to 8 hours of bright sunlight outdoors. By providing about 12 hours of supplemental light inside, you can generally keep

Even this bright windowsill doesn't provide enough light for several succulents in these planters. They have become overstretched, and some have toppled over.

223

Some of these succulents will need to go indoors for the winter.

succulents looking their best. Succulents vary in their photoperiodic requirements for flowering. Some, such as Christmas and other types of holiday cactus (*Schlumbergera* spp.), are triggered to flower under short days, while others need lengthening days and warming temperatures.

The Right Temperature

Succulents perform best with warm-season temperatures between 70°F and 80°F (21–26°C), so normal indoor temperatures will do nicely. Succulents will benefit, however, from a cooling of temperatures in winter, ideally 50°F (10°C) and warmer. Most homes won't be that cool in winter, and plants can tolerate warmer temperatures. A cool garage or basement is a good option if you can move plants during winter months. Some succulents are hardy perennials and can survive cold temperatures, but make sure you research your particular variety's outdoor hardiness if you plan to move plants outdoors.

Water and Feeding

If you are an overwaterer by nature, be prepared to kill a lot of succulents. Plants are naturally adapted to go through very dry cycles, and those fleshy leaves store the water they need to survive. If you keep succulents constantly wet in an indoor environment, they will quickly rot. Always use a loose, well-draining potting mix blended for succulents, and avoid containers that are too large for the existing root system. Allow plants to dry between waterings. In winter, when temperatures are cooler, you can dial back the watering even further.

Most cacti, such as this sea urchin cactus, *Echinopsis*, require very bright light and will not survive indoors without grow lighting.

Succulents tend not to be heavy feeders, so go easy on the fertilizer. During spring and summer use a balanced houseplant fertilizer or liquid humus diluted to one quarter the normal application strength each time you water, which may only be a couple of times per month (or less, depending on the type of plant).

You won't experience many pests on indoor succulents if you care for them properly and they get enough light. Stressed plants, however, can fall victim to spider mites, whiteflies, mealy bugs, and scale. Fungus gnats can be a problem in the soil of any plants. Fungal crown rot, which results from overwatering, is often the biggest killer of indoor succulents.

You can collect and grow many types of succulents indoors. They will grow best if you provide them supplemental lighting.

Echeveria This is a genus of succulent plants, mostly native to Mexico and South America, that forms rosettes of foliage in colors ranging from silver to orange-red. They are beautiful container foliage plants that will also produce flower spikes

from spring through fall. Plants need bright light or they will quickly stretch and go into decline. You can vegetatively propagate plants from leaves and stem cuttings. Some echeverias are long-day plants, while others have more complicated photoperiod responses, such as long days followed by short days (LSDP) or short days followed by long days (SLDP). They are mostly grown for their foliage, so flower induction is rarely a concern.

Euphorbia Euphorbias are sometimes grouped together with cactus plants, as many have thorns. There are many species that vary widely in color, shape, and form; some grow very large and have interesting angular stems. Most have a milky sap that can irritate the skin. Plants need bright light to thrive. You can vegetatively propagate plants from leaves and stem cuttings. Often short-day plants.

Graptopetalum Ghost plant, *Graptopetalum paraguayense*, produces rosettes of thick, beautiful blue-silver foliage. Plants are related to *Echeveria* and will grow under the same bright-light conditions. You can vegetatively propagate plants from leaves and stem cuttings.

Lithops Also known as living stones, lithops are tiny succulents that resemble small smooth stones. They can be tricky to grow indoors because different species may have different environmental needs, but all need bright light. They are highly desirable, especially for collectors who love miniature plants. Plants require a dormant season, when you should stop watering altogether, fall through winter, but different species will have different watering requirements. While most succulents are grown for their foliage, not their blooms, flowering in lithops is a special treat.

Donkey tail spurge,
Euphorbia myrsinites,
offers beautiful pale blue
foliage and yellow flowers.

RIGHT Ghost plant is
impressive in wide, low
containers.

Some species are short-day plants, while others are long-day plants. Research the particular species you have purchased. You can grow plants from seed or division.

Senecio Upright and trailing succulents, often with finger-like or bead-like foliage that has a gray or blue-silver tint. There are many different interesting forms in this succulent genus. Some species are day-neutral, while others may be short-day or long-day plants. Plants need bright light to thrive and, as with any succulent, take care not to overwater, especially in winter months.

LEFT The spiky blue leaves of *Senecio* 'Blue Finger'.

Living stones—don't you just want to pinch them?

CONCLUSION

Growing plants under lights indoors is a rewarding, fascinating pursuit. Just as with any acquired skill, the amount of time and energy you put into learning the art of indoor gardening will influence your success. Plants are living things that respond just as individually as we do to their environment. Don't expect to get everything right the first time. Stick with it and learn from your mistakes. No one is really born with a green thumb—we all have to earn it!

Whether you're growing microgreens on the kitchen counter, harvesting tomatoes from your closet in January, or cultivating orchids in your garage grow tents, there is nothing like the success of feeding yourself or your family from your indoor garden. And getting that prized orchid to finally rebloom? Priceless.

I hope you have enjoyed this book and can use it to begin, and continue, the beautiful experience of growing your own food and flowers. Now go get your hands dirty.

RESOURCES

There are many good sources for grow lighting, supplies, plants, and horticulture education. This list is by no means comprehensive, but it does feature sources I use myself. It's a starting place for you to buy and learn.

You can always visit and contact me online at **lesliehalleck.com** for my Plantgeek Chic blog and gardening information.

On social media:

Instagram and Twitter: @lesliehalleck
Facebook: facebook.com/HalleckHorticultural
Pinterest: pinterest.com/lesliehalleck
LinkedIn: linkedin.com/in/lesliehalleck

GEAR AND SUPPLIES

Apogee Instruments, Inc. manufactures state-of-the-art quantum flux meters for horticultural and aquatic needs. **apogeeinstruments.com**

Arizona East is a grower of stylish succulents you can buy at your local garden retailer; be sure to follow them on Instagram. **arizonaeast.com**

Gardener's Supply Company offers a plethora of gardening supplies for both the indoor and outdoor garden, including small indoor growing systems. **gardeners.com**

Garden Supply Guys, Inc. is a friendly retail indoor growing–supply store with a comprehensive and easy-to-use website for online ordering of grow lamps and supplies. **gardensupplyguys.com**

Homestead Gardens is an independent retail garden center in Maryland that has embraced the Modern Homesteading movement. **homesteadgardens.com**

Opcom Farm designs and manufactures unique indoor growing systems for the DIY home farmer. **opcomfarm.com**

Nui Studio is a design studio in Germany that blends function and form to create inspiring pieces of furniture and unique grow lighting. **nui-studio.com**

Soltech Solutions, Inc. designs and sells attractive LED grow lamps for your indoor living spaces. **soltechsolutionsllc.com**

Sunlight Supply, Inc. manufactures and distributes a wide variety of grow lamps and growing gear for the serious and subtle enthusiast, as well as educational content. **sunlightsupply.com**

7Sensors manufactures self-contained growing units for growing a variety of plants. **7sensors.com**

PLANTS

Baker Creek Heirloom Seeds offers a wide variety of heirloom rare seeds. **rareseeds.com**

Ball Horticultural Company is a world leader in plant development and distribution. **ballhort.com**

California Carnivores offers a broad selection of plants with good availability and supplies. **californiacarnivores.com**

Four Winds Growers is a specialty grower of dwarf citrus trees that can be grown in containers. Note that citrus cannot be shipped to all states. **fourwindsgrowers.com**

Logee's offers up a unique online and in-store selection of unique tropical ornamental and edible plants for the indoor garden. **logees.com**

Seattle Orchid offers online ordering for a wide variety of common and specialty orchids. **seattleorchid.com**

Sustainable Seed Company offers organic and heirloom seeds. **sustainableseedco.com**

Territorial Seed Company offers a wide variety of hybrid and heirloom seeds. **territorialseed.com**

ADDITIONAL LEARNING

The American Orchid Society website offers up plenty of information to get you started on your way to a full orchid addiction. **aos.org**

The Cactus and Succulent Society of America publishes an in-depth journal. **cssainc.org**

Grow Weed Easy offers up educational content and resources to cannabis enthusiasts. **growweedeasy.com**

High Times provides educational information and resources to those interested in growing cannabis indoors. **hightimes.com**

The International Carnivorous Plant Society provides a plethora of information on meat-eating plants. **carnivorousplants.org**

The Urban Jungle Bloggers blog will inspire you with plant styling tips for your living space. **urbanjunglebloggers.com**

The United States Department of Agriculture National Agriculture Library provides information on topics such as home gardening, growing edibles, sustainable agriculture, hydroponics, and aquaponics. **nal.usda.gov**

Be sure to check out the entire **Timber Press** catalog of books for more resources on specific types of plants and horticultural information. **timberpress.com**

REFERENCES
AND BIBLIOGRAPHY

Apogee Instruments, Inc. PPFD to LUX Conversion Tables. apogeeinstruments.com/unit-conversions.

Ball, V. 1991. *Ball Redbook*. West Chicago: Geo. J. Ball Publishing.

Baldwin, D. L. 2013. *Succulents Simplified: Growing, Designing, and Crafting with 100 Easy-Care Varieties*. Portland, Oregon: Timber Press.

Capon, B. 1990. *Botany for Gardeners: An Introduction and Guide*. Portland, Oregon: Timber Press.

Finical, L. M. 1998. *The Effects of Plant Age, Photoperiod, Vernalization, and Temperature on the Growth and Development of Aquilegia and Gaura* (Master's thesis). East Lansing, Michigan: Michigan State University Press.

Hewitt-Cooper, N. 2016. *Carnivorous Plants: Gardening with Extraordinary Botanicals*. Portland, Oregon: Timber Press.

Janick, J. 1986. *Horticultural Science*. New York: W.H. Freeman and Company.

Josifovic, I. and J. De Graaff. 2016. *Urban Jungle*. Munich: Verlag Georg D.W. Callwey & Co. KG.

Kozai, T., G. Niu, and M. Takagaki. 2016. *Plant Factory: An Indoor Vertical Farming System for Efficient Quality Food Production*. New York: Elsevier, Inc.

Lopez, R.G. and E. S. Runkle. 2004. The Flowering of Orchids. *Orchids* (March): 196–203.

Morgan, L. 2013. Daily Light Integral (DLI) and Greenhouse Tomato Production. *The Tomato Magazine* 17(4): 10–15.

National Geographic. 2008. *Edible: An Illustrated Guide to the World's Food Plants*. Lane Cove, Australia: Global Book Publishing.

Rosenthal, E. 2010. *Marijuana Grower's Handbook: Your Complete Guide for Medical and Personal Marijuana Cultivation*. Oakland, California: Quick American Publishing.

Smith, E. C. 2009. *The Vegetable Gardener's Bible*. North Adams, Massachusetts: Storey Publishing.

Taiz, L. and E. Zeiger. 1991. *Plant Physiology*. Redwood City, California: Benjamin Cummings Publishing Co., Inc.

Thimijan, R. W. and R. D. Heins. 1982. Photometric, Radiometric, and Quantum Light Units of Measure: A Review of Procedures for Interconversion. *HortScience* 18: 818–822.

Torres, A. P. and R. G. Lopez. 2011. Commercial greenhouse production. *Purdue Extension*.

Warner, R. M. 2006. Supplemental Lighting on Bedding Plants: Making it Work for You. *OFA Bulletin* 899.

Wang, Y. et al. 2013. Temperature and Photoperiod Impact Orchid Spiking. *Greenhouse Grower* 31: 10–48.

PHOTOGRAPHY AND ILLUSTRATION CREDITS

7Sensors, page 53 bottom left

Alamy Stock Photo
Beat Bieler, page 183
Glasshouse Images, page 160
Graham Corney, page 169
Leila Cutler, page 32 top
RBflora, page 168
The Picture Pantry, page 178
Valentyn Volkov, page 172 bottom left
Zigzag Mountain Art, page 161

Apogee Instruments / Elisa Wilde, page 44

Arizona East, pages 8, 221 bottom left and bottom right

Ball Horticultural Company, pages 30 bottom, 37 top, 100, 153 right, 176, 193, 196, 223 bottom, and 225 right

Garden Supply Guys, pages 20, 28 top left and top right, 70 top, 71 top left, and 148

Homestead Gardens, pages 26, 28 bottom left, 53 bottom right, 78, 89 left, 90, 149 bottom left, 149 bottom right, and 199 bottom right

iStock.com
bhofack2, page 173 top
Denira777, page 177
dvulikaia, page 184
HandmadePictures, page 156
jc_design, page 188
JillLang, page 31
Marc_Espolet, page 45
MSPhotographic, page 172 top left
Neustockimages, page 157
robertcicchetti, page 13

Nui Studio, page 77

Pot Pots / Dan Heims, pages 27, 33, 163, 164, and 165

SolTech Solutions, LLC, pages 46 left, 75, 203, and 208 bottom

Sunlight Supply, Inc., pages 25 top, 34, 42 top left, 53 top, 57 top left, bottom left, and bottom right, 62 bottom, 64 bottom left, 65 top, 66 right, 67, 68, 70 bottom, 71 top right and bottom, 74 bottom, 76, 82, 83 top left and bottom, 93 bottom, 128, 131 bottom, 143 top left, and 149 top right

All other photos are by the author.
All illustrations are by Arthur Mount.

ACKNOWLEDGMENTS

After college, I secured a spot as a graduate student at Michigan State University in the floriculture department. I spent my time there under the tutelage of Dr. Art Cameron, Dr. Royal Heins, and the late Dr. Will Carlson. They gave me a run for my money when it came to learning the science of light, and I thank them for it. (I'm pretty sure I gave them a good run in turn.)

These days, running my green-industry consulting company keeps me busy. Writing a book on top of my day job was no light work. I must thank my two assistants for their help on this project: Elizabeth Krause for wrangling my words and Nikki Rosen for wrangling resources. Mark Aufforth is much appreciated for his "I can build anything" skills. I'd also like to thank those who contributed images and gear, listed in the resources, especially Garden Media Group and Sunlight Supply. Thanks to my parents for letting me take sequestered writing sabbaticals at the family house. Of course, I extend special gratitude to Tom Fischer and the rest of the team at Timber Press. It takes a village. Most important, I thank my ever-patient and understanding husband, Sean Halleck, for flying solo for a year's worth of weekends (complaint free) so I could focus on writing this book. Go team.

INDEX

biodegradable containers, 129
bio-security, 108, 110
black-eyed pea, 159
bladderworts, 207
blazing star, 126
Bletilla, 213
blood orange, 173
blood sorrel, 189
blue chives (Siberian), 169
blue vs. red light, 25–26, 44
boat orchid, 219–220
bok choy, 111
bolting, 33, 94
bonsai, 202–207
borage, 125, 192
Botrytis, 115
bougainvillea / *Bougainvillea*, 204
boxwood, 204
Brassia, 213
Brassica oleracea, 177
Brassica oleracea var. *alboglabra*, 161
Brassica oleracea var. *italica*, 161
Brassica rapa, 161
broad beans, 136
broccoli, 94, 135, 161–162
  'Cruiser', 162
  'Green Comet', 162
  'Green Goliath', 162
  'Packman', 162
  'Premium Crop', 162
  'Romanesco Italia', 162
  'Summer Purple', 162
bronze fennel, 176
brugmansia / *Brugmansia*, 85, 140
Brussels sprouts, 162, 178
Bulbophyllum, 213
bulbs, conversion, 68
bulbs, flower, 97, 98, 99, 115
butterfly weed seeds, 137
butterworts, 207
Buxus, 204

C

cabbage, 135, 178
cacti, 220, 221, 223, 224
calabrese, 161–162
calamondin, 173

calculations and conversions
  CFM, 104–105
  days-to-harvest, 136
  DIF, 96–97
  DLI, 45–48, 51
  lumens, 48–51
  number of lamps needed, 78–80
  PPFD / lux, 48–51
  watts per square foot, 78
calendula, 35
cannabis
  CO_2 levels, 106–107
  flowering, 26, 27, 33, 164
  light requirements, 54,
    68, 164–165
  pests and diseases, 111, 112, 113,
    115, 167
  photoperiod, 34, 35, 163–164
  plant profile, 163–167
  temperatures, 152, 166
Cannabis indica, 163, 166
Cannabis ruderalis, 164
Cannabis sativa, 34, 163, 166
Cape sundew, 206
capillary mats, 132, 133
Capsicum annuum, 186
Capsicum baccatum, 186
Capsicum chinense, 186
Capsicum frutescens, 186
Capsicum pubescens, 186
Carmona retusa, 204
carnivorous plants, 205–207
carrots, 35, 136, 167–168
  'Chantenay', 168
  'Danvers Half Long', 168
  'Imperator', 168
  'Little Finger', 168
  'Thumbelina', 168
caterpillars, 111, 118
catmint, 182
catnip, 182
catteleya / *Cattleya*, 213, 216
cauliflower, 162
cedar, 204
Cedrus, 204
celery, 135, 170
centranthus, 35
Cephalotus follicularis, 207

CFL (compact fluorescent) lamps,
  51, 59–60, 61, 62, 68
CFM (cubic foot per minute)
  ratings, 104–105
chard, 35, 111, 178
cherry, 98, 204
chervil, 170
chicory, 35
Chinese broccoli, 161–162
Chinese cabbage, 179
Chinese elm, 204
Chinese parsley, 169–170
chitting (or greensprouting), 124
chives, 168–169
chlorophyll, 14–17, 20, 146
chloroplasts, 12, 14, 20, 25
chlorosis, 146
Christmas cactus, 224
chrysanthemums, 31
cilantro, 35, 169–170
  'Leisure', 170
  'Long Standing', 170
  'Slow Bolt', 170
×*Citrofortunella microcarpa*, 173
Citrus aurantifolia, 171
citrus canker, 116
citrus / *Citrus*
  cuttings, 143
  pests and diseases, 109, 112, 113,
    115, 116, 118
  plant profile, 171–173
Citrus hystrix, 171
Citrus latifolia, 171
Citrus limon, 171
Citrus reticulata, 171
Citrus ×*sinensis*, 171
Cladosporium, 116
clay pellets, 131
cleome, 35
cloning. *See* cuttings
CMH (ceramic metal halide) lamp,
  27, 46, 62–65, 90
CO_2 (carbon dioxide) levels, 106–107
coir (coco fiber), 127–128, 140
cole crops, 111
collards, 94
columbine, 98

fuchsia, 35
Fukien tea tree, 204
full-spectrum labels, 38–39, 76–77
full-spectrum light, 27–29, 76–77
full-spectrum sensors, 44–45
fungal diseases, 104, 118
fungus gnats, 110, 111, 118

G
Ganoderma, 114
garlic, 34, 35, 98, 99, 100
garlic chives, 169
gazania, 35
germination
    days-to-harvest, 135–136
    dormancy and, 126
    light and dark requirements,
        23, 24, 122
    rates, 125, 131, 137, 157, 188
    temperatures, 96–97, 132–134
gesneriads, 200, 201
ghost plant, 225, 226
giant Siberian chives, 169
globe amaranth, 35
grafting, 26
grandiflora petunia, 35
grapefruit, 173
grapes, 115
Graptopetalum, 225
Graptopetalum paraguayense, 225
grasses, 111
gray mold, 115
green fennel, 176
green light, 20–21
greens
    Asian greens, 179
    collard greens, 94, 178
    leaf miners, 112
    light requirements, 39, 52, 53, 54,
        57, 60, 62, 90
    microgreens, 25, 34, 52, 103,
        133, 180–181
    mustard greens, 179
    photoperiod, 32
    temperatures and flavor, 94
grouping plants with similar needs,
    34, 152, 205

growing container sizes and spacing,
    144–145, 155. *See also plant*
    profiles
growing media, 127–128, 140. *See*
    also plant profiles
grow tents, 88–93, 96, 104, 107

H
hardneck garlic, 98, 99, 100
Haworthia, 222
heat mats, 127, 134
hens and chicks, 221
herbs
    growing spaces, 90, 92, 149, 208
    light requirements, 52, 53, 90
    pests and diseases, 111, 112,
        115, 116
HID (high-intensity discharge)
    lamps, 60–62, 76
HO (high-output) T5 fluorescent
    lamps. *See* fluorescent lamps
Hoya, 201
HPS (high-pressure sodium) lamps,
    27, 39, 61–62, 65–66, 81,
    85, 90, 92
humidity, 101–102, 104
humidity domes, 102, 103, 132, 133
hyacinth, 98
hyacinth bean vine, 35
hybrid petunias, 35
hybrid tulips, 98
Hydrangea macrophylla, 97
hydrangeas, 97
hydroponic growing systems, 45, 110,
    114, 127, 131, 143, 148–150
hydroton, 131

I
Ilex crenata, 204
incandescent lamps, 54–55
induction and plasma lighting. *See*
    LED lamps; LEP lamps; mag-
    netic induction lamps
inputs, 17, 39–40, 52, 78
insecticidal soaps, 117
insect predators, 118
Iochroma, 201

IR (infrared radiation), 22, 23, 26

J
jade bonsai, 204
jade plants, 204, 223
Japanese elm, 204
Japanese maple, 204
juniper, 204
Juniperus, 204

K
kalanchoes, 223
kale, 94, 95, 135, 177–178
    'Dinosaur', 178
    'Vates Blue Curled', 178
    'Winterbor', 178
Kelvin ratings, 38–39
Kohleria, 201
kumquat, 173

L
Lactuca sativa, 178
ladybug larvae, 118
lady slipper orchid, 218–219
lamps
    cost considerations, 56–57, 61, 62,
        68, 69, 71, 80
    key factors when choosing, 52
    placement, 46, 48, 83–85, 122–124
    safety, 68
    shop lights, 52
    for starting seedlings, 124
    switching between, 26, 83
    See also specific lamp types
lamp timers, 34
late blight, 115
lavender, 194
leaf miners, 112
LEC (Light Emitting Ceramic)
    lamps, 62–63, 64, 65, 78
LED lamps
    configurations in a home setting,
        69, 73, 77
    full-spectrum label, 76–77
    HID LEDs, 76
    light spectrum, 20, 25,
        26–27, 70–71

Leslie F. Halleck is a certified professional horticulturist (ASHS) who has spent her 25+ year career hybridizing horticulture science with home gardening consumer needs. Halleck earned a B.S. in Biology/Botany from The University of North Texas and an M.S. in Horticulture from Michigan State University. The focus of her graduate research was greenhouse plant production using environmental controls such as lighting, temperature, photoperiod and vernalization. Halleck's professional experience is well-rounded, with time spent in field research, public gardens, landscaping, garden writing, garden center retail and horticulture consulting. At the end of 2012 Halleck devoted herself full-time to running her company, Halleck Horticultural, LLC, a green industry consulting and marketing agency. Halleck's previous positions include Director of Horticulture Research at The Dallas Arboretum and General Manager for North Haven Gardens (IGC) in Dallas, TX. Halleck is now a regular feature on the professional speaking and industry publication circuit, but she continues to offer up common sense gardening advice and hands-on learning to home gardeners via her Plantgeek Chic blog, public workshops and consumer publications. During her career, Halleck has written hundreds of articles for local, regional and national publications as well as taught countless gardening programs for the home flower gardener, edible enthusiast and backyard farmer.